I0879649

SLATE ROOFS
1926

ISBN: 978-1-7336035-0-8

Library of Congress Control Number: 2020901720

Compiled, edited, and designed by Joseph C. Jenkins.
Published by Joseph Jenkins Inc.
143 Forest Ln, Grove City,PA 16127 USA Ph: 814-786-9085

Printed in the USA

JosephJenkins.com

CONTENTS

Slate Roofs 1926:

Addendum:

From the Slate Roof Bible 3rd Edition (2016):

On the cover: Renwick Chapel, Oak Hill Cemetery, Georgetown DC. Photo courtesy of Chuck Wagner, Wagner Roofing

FOREWORD 2020

It is with great pleasure, honor, and humility that I bring back into print the classic "Slate Roofs," published in January 1926 by the then nascent "National Slate Association" (page 84). Although I adapted the title to be more compatible with a 21st century readership, and I made some corrections and stylistic changes, almost all the information is presented exactly as in the original publication, even anachronistic references. For example, in my 52 years in the slate roofing industry, I never heard of or saw a "flat slate roof" other than in this publication (page 64). However, this book contains a wealth of information on which today's modern slate roofing industry is based. It therefore remains a valuable addition to the library of anyone interested in roofing, construction, architecture, and history. Included at the end of this publication are 21st century guidelines for slate roof installation, repair, and restoration as published by the Slate Roofing Contractors Association of North America, Inc., as well as excerpts from the 2016 Slate Roof Bible 3rd Edition.

Joseph Jenkins, author and publisher of "The Slate Roof Bible" January 2020

[Note: References in this publication to the "National Slate Association" formed in 1922, are not in any way referring to or associated with the "National Slate Association" that formed in 2002.]

FOREWORD

THIS publication is the result of the active co-operation of architects, slate producers and distributers, roofing contractors, governmental departments, manufacturers of accessories and many individual authorities. To all of these The National Slate Association expresses its sincere appreciation of their support and interest.

The essence of all this information supplied to the Association and to the Structural Service Bureau and its consulting architect, has been incorporated in the book to the fullest extent possible.

In offering this handbook on slate roofs the National Slate Association has been actuated by the desire to supply reliable and definite information on slate and its proper application. It is hoped that this publication will make for the more extensive and appropriate use of slate as a practical, economical and attractive roofing medium besides supplying information on formation and quarrying to users of this invaluable and most unique stone.

For the further benefit of the industry and of all specifiers and users of roofing slate, recommendations concerning the arrangement and contents, which may make later editions of this publication of even more practical use or widen its scope, will be received with appreciation by

NATIONAL SLATE ASSOCIATION

Philadelphia, Pa.

Each illustration in this book has been carefully selected from the large number of photographs available for the sole purpose of making clear some particular point in the text. For this reason the names of the Architect, Slate Producer and Contractor have been omitted, but can be furnished by the Association if desired.

First Edition
January, 1926

This is the Second of a Series of Informative Literature on Slate and Its Uses.
The First—"Charm of Slate Floors and Walks"—Published January, 1925.

"Slate Roofs"
Price, $1.50

A Picturesque Slate Roof — *H. T. Lindeberg, Architect*

DIGEST OF CONTENTS

(For Complete Index See Page 83)

COMPLETE AUTHORITATIVE AND PRACTICAL DATA, DRAWINGS AND SPECIFICATIONS ON ROOFING WITH SLATE

THE practical purpose of any roof is to protect the interior, the contents and the occupants of a building from rain, snow, heat and cold. In addition to these practical considerations, the roof should add to the appearance and character of the building, and the passing of time should only enhance its beauty and add to its intrinsic value.

The period of usefulness of the roof will depend upon the resistance of the roofing material itself and the materials with which it is laid to the action of the elements. To be permanent, the roofing material must be unaffected by the action of water, climatic changes and gaseous fumes in the air, and must also be fireproof. For economy, it should require no other material to preserve it. However, even though the material possesses these characteristics, a permanent roof will not be secured unless it is properly laid and its fastenings are selected for the same enduring qualities.

The permanent and fireproof qualities of slate make it eminently suitable for either sloping or flat roofs. In no other roofing material will be found so many characteristics combining to offer such an alluring and indefinable variety in color, texture and line.

This is easily understood when it is recalled that slate is a stone, formed by nature to serve the diversified uses of man. It requires no admixture of materials, domestic or foreign; it needs no heat to form it and no process to manufacture it other than the handiwork of extracting it from the ground in blocks, splitting and trimming them to the desired size and thickness. Moreover, the first cost of slate is not as great as that of many fabricated products, and with its practically negligible maintenance, it becomes the least expensive roof covering available when service and appearance are taken into account. Besides annual savings in insurance rates most slates when removed from old roofs with care are as good as new.

NATIONAL SLATE ASSOCIATION.
1926

SLATE ROOFS 1926

A Handbook of Data on the Constructing and Laying of All Types of Slate Roofs

Slate roofs, the natural stone coverings, may be conveniently classified as:

Standard Slate Roofs.
Textural Slate Roofs.
Graduated Slate Roofs.
Flat Slate Roofs.

The following briefly describes each type of roof. More detailed information and a specification for each will be found on pages 48 to 65.

Standard Roofs

Standard slate roofs are those composed of slate approximately 3/16″ thick (Commercial Standard Slate), of one uniform standard length and width, having square tails or butts laid to a line. Slate of this type is commonly obtainable in the basic slate colors. Standard roofs are suitable for any building where a permanent roofing material is desired at a minimum cost. It differs from other slate roofs only in characteristics affecting the texture or appearance of the roof, through the shape and thickness of the individual units. If desired, the butts or corners may be trimmed to give a hexagonal, diamond or "Gothic" pattern for all or part of the roof, as, for instance, on a church spire. Standard Roofs are sometimes varied by laying two or more sizes (lengths and widths) of Commercial Standard Slate on the same area. For specification see page 52.

Figure 1. Standard Roof

Figure 2. Textural Roof

Textural Roofs

The term "Textural" is used to designate those slates usually of rougher texture than the Standard, with uneven tails or butts and with variations of thickness or size. In general, this term is not applied to slate over ⅜″ in thickness. Varying shades are frequently used to enhance the color effect, which, with the characteristics just mentioned, add interest in line and texture to the roof design. In addition to the basic colors of the commercial grades, accidental colorings of bronze, orange, etc., may also be used in limited quantities. For specification see page 55.

Graduated Roofs

The graduated roof combines the artistic features of the Textural Slate roof with additional variations in thickness, size and exposure. The slates are so arranged on the roof that the thickest and longest occur at the eaves and gradually diminish in size and thickness until the ridges are reached. Slates for roofs of this type can be obtained in any combination of thicknesses from 3/16″ to 1½″ and heavier when especially desired.

The graduated slate roof presents many opportunities for variation and offers excellent possibilities for interesting treatment. The fact that it is especially designed to harmonize with the general character of the building of which it becomes a part or to meet exacting requirements of construction causes it to be frequently termed the custom-made roof of the industry. Many producers and distributors maintain special design staffs to assist architects in securing the most suitable and satisfactory graduated slate roofs. These services are freely offered to designers in the interest of better and more harmonious slate roofs. For specification see page 58.

Flat Roofs

Flat roofs offer a wide field for roofing slate, and are so designated whether or not they are used for "promenade" purposes.

Slates of any thickness may be used in place of the slag or gravel as a surfacing material for the usual built-up type of roof. The body, weight and enduring qualities of slate make it highly desirable as a protection to the waterproofing beneath the surface, whether or not it is subject to traffic. Only the thicker slates are used on promenades. For specificational data see page 62.

Figure 3. Graduated Roof

Figure 4. Flat Roof

SLATE

Characteristics

The principal difference between slate and other stones is the natural slaty cleavage of the former, which permits it to be more easily split in one direction than in others. A second direction of fracture or "scallop," usually at right angles to the slaty cleavage, is called the "grain." Roofing slates are commonly split so that the length of the slate runs with the grain.

Slate quarried for roofing stock is of dense, sound rock, exceedingly tough and durable. Slate, like any other stone, becomes harder and tougher upon exposure than when first quarried. It is practically non-absorptive, tests on Pennsylvania slate showing a porosity of 0.15 to 0.4 percent.

The nature of the surface after splitting is dependent upon the character of the rock from which it is quarried. Many slates split to a smooth, practically even and uniform surface, while others are somewhat rough and uneven. As a result, a wide range of surface effects is available for the finished roof.

Slate from certain localities contains comparatively narrow bands of rock differing to various degrees in chemical composition and color from the main body of the stone. These bands are called "ribbons." Ribbons containing no injurious constituents and not of undesirable color are not objectionable. Slates of this type, when trimmed so that the ribbons are eliminated, are known as "clear slate." Slates which contain sound ribbons are sold as "ribbon stock."

Color

The color of slate is determined by its chemical and mineralogical composition. Since these factors differ in various localities, it is possible to obtain roofing slates in a variety of colors and shades.

It is truly remarkable to find a natural product possessing, in addition to the other qualities before mentioned, such unlimited possibilities in color effect. The use of slate for roofing makes it possible to obtain a surface of uniform or contrasting colors in cold or warm values. Moreover, if the design of the building requires a roof of one general color, it may be graded up or down the slope from dark to light as desired.

Figure 5. Skulping Large Slate Slab

To relieve the monotony of a flat uniform body color, various shades of the same color may be used to provide an interesting variation up and down or across the roof or interspersed throughout. A low-eaved, prominent roof surface may require a quiet or contrasted blending of the autumnal colors of nature to cause the structure to take its place in its intimate surroundings. Slate not only permits a roof of permanent color, but by the selection of "weathering" slate, one may take advantage of the mellowing effect of age and weather. These and all the steps between these extremes are ever ready at the designer's bidding in the field of slate.

Upon exposure to the weather, all slate is changed slightly in color. The extent of this color change varies with different slate beds, being barely perceptible in certain slates. Those slates in which the color changes but slightly are classed as "permanent" or "unfading." Those in which the final result is more marked and varied are known as "weathering" slates.

Weathering is the natural result of exposure to the weather of the coloring minerals in the slate. Where color is an essential consideration, architects and owners should take this characteristic into account. The quarry operators know from experience the probable nature and extent of the changes in the original color, although different quarries in the same state, and often in the same locality, may produce a wide range of colors in both "permanent" and "weathering" slates.

To take the fullest advantage of the various effects obtainable, the source of the slate, as well as its ultimate color effect, should be known.

Color Nomenclature

For the purpose of utilizing the basic natural colors of roofing slate available in large quantities for general usage, The Division of Simplified Practice of the Department of Commerce has recommended the following color nomenclature:

Basic Slate Colors

Black	Grey	Purple	Green
Blue Black	Blue Grey	Mottled Purple and Green	Red

These color designations should be preceded by the word "unfading" or "weathering," according to the ultimate color effect that may be desired.

For special treatment roofs certain quarries supply other colors and combinations of colors.

Many requests were received for inclusion in this book of plates reproducing slate colors. After giving full consideration to the various phases and possible misconception which surround color reproduction, however, it was decided not to include such plates. Some individual producers have prepared plates but much confusion has been caused by purchasers believing and insisting that each slate on their roof should be an exact match in tone and marking with the one shown in the plate. This, of course, is impracticable when dealing with the eye-resting tones nature has bestowed upon slate with such pleasing variations and gradations even in the same slate and between slates. Samples of slates or color plates furnished by producers can only be taken as indicative of the color tones of certain slates.

All slates would be grey, the commercially known slate color were it not for the coloring minerals present, for example, chlorite in the greens, hematite in the purples and hematite and iron oxide in the red. (See page 75 for table of Mineral Constituents of Slate.)

AMERICAN ROOFING SLATES

Available in Commercial Quantities in All Thicknesses Unless Otherwise Noted

COLOR DESIGNATION	WHERE QUARRIED	BRIEF DESCRIPTION
Blue-Grey	Lehigh-Northampton Counties Pennsylvania. Veins extend through entire slate zone.	Probably the best known, most widely used and most accessible of the basic slate colors. Sales records substantiate this fact as the largest shipments of commercial standard slates are made from this district. Production capacity of quarries now in operation assure architects and the public of unlimited supply of this popular color. Just as it is the more generally recognized slate color so is it also the base in cost or price tables of distributors and roofing contractors and nearly all of the latter carry it in stock. Because its blue grey tones harmonize so well with nature's other handiwork "The Blue Sky" may explain why it is so frequently selected for the "roof over his head" of every man and by architects in carrying out their effects desired. Acres of these slates have been specified and shelter vast numbers of the state, federal and other public and private institutional buildings. To more nearly portray the true color characteristics of these Pennsylvania slates they have been designated "Blue-Grey." Some architects and older roofers may still continue to refer to them as Pennsylvania black, but to avoid confusion and chance of receiving "Black-Bed" slates (Note 2) architects, builders, roofers and others are cautioned to use this classification "Blue-Grey" as the color designation. Some splendid color combinations have been evolved by working together the different shades of these slates, or blending them with the weathering greens of Vermont or with the clear purples where a dark underlying tone is desired.
Blue-Grey	Near Esmont, Virginia and the Arvonia Belt of Buckingham and Fluvanna Counties, Virginia.	A tough durable Slate and one which makes an especially beautiful roof. This slate is notable for a peculiar luster giving very attractive lights and shadows on the roof.
Unfading Grey	Northampton County, Pennsylvania.	Very popular with architects because of their soft grey tones which harmonize so well with almost any type of wall surface. Unusual transverse strength. Characterized by extremely large quartz crystals and sericite flakes.
Grey	Western part Vermont, Washington County, New York. Veins extend through entire slate zone.	Considerably lighter in tone than Blue-Grey slates of Pennsylvania. In combination with clear purples provides an excellent mixture where dark value underlying tone is preferable to a singleness of color. Some weather and some are unfading.
Grey-Black	Western part Vermont, Washington County, New York. Veins extend through entire slate zone.	Available in light and dark shades in either plain or mottled effect. Some are unfading and some weather. Most of the grey slates are mottled by streaks of darker grey and for this reason are often used to advantage with other slates.
Blue Black—Hard Vein	Chapman Quarries, Pa., and adjacent part of Northampton County, Pa.	Exceptionally hard durable slate, having one or more hard veins running across the slate. These veins produce a texture and color effect very much desired. The color tone grows darker with exposure to elements.
Unfading Black	Peach Bottom, York County, Pa., and adjacent part of Maryland. Piscatsquis County, Maine.	Peach Bottom Slate generally is heavier than many other slates and has a rougher appearance, and it is of the unfading color group. Markedly graphitic. First American roofing slates quarried. Monson roofing slate, quarried in the town of Monson, Maine has long been known as one of the strongest and best roofing slates produced in this country. It is an unfading black slate having a slight lustre, available only in standard thickness of $\frac{3}{16}$". Nothing over this thickness is produced except on occasional and special orders.

Unfading Green	Western part Vermont, Washington County, New York. Veins extend through entire slate zone.	Available in light or dark shades. Whatever fading or weathering takes place is so uniform that a permanently green roof color is assured.
Weathering Green	Western part Vermont, Washington County, New York. Veins extend through entire slate zone.	Not quarried in commercial quantities in northern part of producing zone. Available in light and dark shades. Exceedingly strong and tough, of excellent texture. Upon exposure to the elements some of these "weather" and are transformed into soft tones of brown, buff and gray while others retain their original shade. This weathering action of the elements serves to beautify a roof of this material and in no way affects the durability of the slate. They harmonize with almost any style of architecture and are used on buildings of every description. Quarries now in operation assure architects and the public an unlimited supply of this popular and most widely known and used of Vermont basic slate colors. Carried in stock by many roofing contractors.
Unfading Purple	Western part Vermont, Washington County, New York. Veins extend through entire slate zone.	Because of tremendous demand for purples in graduated roofs of assorted colors, it is now difficult to furnish clear purples in large quantities.
Unfading Mottled Purple and Green	Western part Vermont, Washington County, New York. Veins extend through entire slate zone.	No two slates have exactly the same marking or colors. They consist of slightly varying shades of purples, available in light and dark tones, sometimes almost clear, frequently clouded with spots or traces of green, others with light and dark clouds of purple, while occasionally there are some having only a purplish tint that might other wise pass for Green Slate.
Variegated Purple	Western part Vermont, Washington County, New York. Veins extend through entire slate zone.	These slates are of a predominating purple tone, some having spots or streaks of green of varying size. In a small percentage, too, green may predominate with only tints of purple. As in the case of other weathering Vermont slates, a percentage change upon exposure to harmonious shades of brown, while others present an unchanged appearance.
Unfading Red	Washington County, New York only.	The only place in the U. S. so far where red roofing slate has been found in deposits of sufficient size to insure profitable operations. Excellent in color, cleavage and strength. The color becomes more pleasing with age. A study of the price lists of slate distributors shows this color as the most expensive.
Freaks	There are delightful combinations or variations in these colors available only for roofs of special treatment. Because of their color they have been termed freak slates. They are just what the name implies—freak slates from rock formations centuries old, of a character that will not permit splitting under ¼″ in thickness and from this up to 2″ for architectural roof purposes. A great array of colors are available, comprising Opals, Bronzes, Buffs and Browns, and others so varied and unique that when displayed en masse one is reminded of a beautiful tapestry. They add charming contrasts and character to a roof. Some of them may not be true slate in every quality that has won for this stone its centuries of sheltering utility, but even should they have no salvage value, they have so enhanced the beauty of the building they have adorned as to fully repay the investment.	

Notes:

(1) Slates from Pennsylvania, especially in a graduated roof, are being widely used to give a wonderfully pleasing color effect by calling for a mixture of blue-black, blue-grey and grey. If so stipulated, the surface or texture may also be had rough and knarled as well as the smooth or standard finish.

(2) One other grade known as "Black Bed" is quarried occasionally in Pennsylvania. Due to an excess content of carbonaceous impurities experience has shown that after many years exposure these slates weather out and some slight discoloration and surface deterioration takes place. The members of this Association warn against the use of such slates except on temporary buildings or isolated rural districts. Through research and the strenuous efforts of the members of the Association only small quantities of such roofing slates are now available as this rock is being diverted into school slates, bulletin blackboards and other uses for which it is ideal. Considering its centuries of sheltering utilization, roofing slate presents a protection record difficult to equal and few failures of slate roofs recorded even after many years of service have chiefly been traced back to this "Black Bed" slate or other slates of high impurity content.

(3) The demand for more careful selection of color tones in furnishing of architectural roofs has led to a more thorough analysis and inventory of the supply. Special characteristics of certain beds or veins in different or even the same quarries produce graduations of weathering tones which require utmost knowledge by roofer of intimate color effect desired on a roof of special treatment so that slates of this class, when used, may be obtained from a quarry which will meet the requirements as nearly as it is humanly possible to be predetermined. Inventive genius some time may eliminate the uncertainty of forecasting or knowing exactly the ultimate color to which slates may weather. The U. S. Bureaus of Standards and Mines and the research engineers of the National Slate Association are working on the problem.

American Roofing Slate Deposits Now Being Quarried

Active roofing slate quarrying in the United States is confined to the states of Maine, Maryland, New York, Pennsylvania, Vermont and Virginia, the chief production of Pennsylvania is from the Lehigh district, including parts of Northampton and Lehigh Counties. The Peach Bottom district extends from York and Lancaster Counties, Pennsylvania, across the line into Harford County, Maryland. (See Map, page 84.)

The active Vermont district lies in Bennington and Rutland Counties, and extends into Washington County, New York. The Maine slate deposits occur in Piscataquis County, about the center of the state. Virginia operations are now conducted only on the Arvonia belt of Buckingham and Fluvanna Counties and near Esmont.

A number of other deposits have been worked in the past and may be worked again, but are now idle; others have not yet been developed. A few others are being worked now exclusively for crushed and ground slate.

In Canada slate has been quarried chiefly in Richmond County, Quebec, though some slate has been produced in Nova Scotia and in British Columbia. Slates have been obtained from the shores of the Bay of Islands and Trinity Bay, Newfoundland. No Canadian quarries were in operation at time of publication of this book.

Grading

Practically all producers have their own trade names and "grades" for slate for "Textural" and "Graduated" roofs and the distinguishing features of each should be familiar to the architect or owner before specifying.

The National Slate Association has on file a complete list of registered trade names of the various slates.

With respect to the characteristics of slate, which have their effect upon grading, Dr. Oliver Bowles, Mineral Technologist of the U. S. Bureau of Mines says, in "The Characteristics of Slate" paper delivered before the American Society for Testing Materials, June, 1923:

> "Slate is of medium hardness, very fine grained of low porosity, great strength and consists essentially of insoluble and stable minerals that will withstand weathering for hundreds of years. Some slate in Pennsylvania contains ribbons which consist of narrow original beds usually containing carbon, and darker in color than the body. There is tendency for some ribbons to contain an excessive amount of the less resistant minerals and they should not appear on exposed surfaces."

Some Pennsylvania slate contains ribbons and the output of some quarries in this district is divided into two classifications known as "Clear" and "Ribbon."

The characteristics which are commonly accepted as affecting the appearance of the slate on the roof namely the surface, straightness, condition of the corners and thickness are used to determine the "Classification" or so-called "Grade" into which the quarries divide their product.

Figure 6. Smooth and Rough Slates

Commercial Standard Means Properly Graded

The classification below applies only to slate less than ¼″ in thickness for "Standard" roofs.

When specifying "commercial standard roofing slate" it is unnecessary to further cover the essential characteristics or grading points to be considered by slate inspectors in selecting and piling the slate in the storage yards at the quarries. The term "commercial standard" embodies certain grading standards which govern the selection of slates for shipment and are as much a part of the process of preparing the slate slabs for roofing purposes as the splitting or any other operation.

It is to be regretted that it is impracticable to have one standard for all parts of the country. It will be realized, however, that this is impossible due to some slates containing ribbons while others are clear, some having a rough surface, others a smooth surface, and certain other distinguishing features which must be given consideration.

In the past many architects have thought it necessary to specify roofing slate by the name of a town or by designated directions from certain towns. That this is too restrictive and unnecessary is apparent when it is realized that any particular color or kind of slate veins may extend through an entire region.

A Number Two Slate comes from the same bed as the Number One and is only so classified because of surface characteristics. A knot or knurl or rougher texture on surface of one slate of a pile split from the same slab does not change its mineral constituents. It is not a manufactured or artificial product which can be varied by formula or human avarice to cut down on any expensive ingredient.

Architects and others may rest assured that their specification of "Commercial Standard roofing slates in accordance with the grading standards of National Slate Association" will obtain all the essential characteristics and quality as though they were to prescribe all the following details of grading standards used at the quarries.

Lehigh-Northampton District, Pennsylvania

There are two grades of Commercial Standard Roofing Slates produced in Lehigh-Northampton District, Pennsylvania, viz:

Number 1—Clear Dark blue, blue-grey, or grey in color, uniform throughout.

Number 1—Ribbon Similar in all respects to No. 1 Clear except for the presence of one or more bands or "ribbons" which are not exposed when on roof.

The following grading rules apply to each of the above grades:

Surface Reasonably smooth straight cleavage full length of slate both front and back. The maximum bend shall not exceed ¼″ in lengths up to 16″, nor exceed ⅜″ in lengths from 16″ to 24″.

Texture Shall be free from knots or knurls that in any way interfere with the safe conveyance or the laying of the slate on the roof.

Corners Reasonably full corners on exposed ends. No broken corners on covered ends that would sacrifice nailing strength, or the laying of a water tight slate roof.

Weight 625 to 750 pounds per square depending on quarry and color.

Thickness Approximately $\frac{3}{16}$″.

In a few Pennsylvania quarries they are also producing two other grades:

Medium Same as No. 1 Clear except that texture is somewhat rougher—less uniform in thickness. Weight 700 to 725 pounds to square.

Number 2—Ribbon Same as No. 1 Ribbon except texture is somewhat rougher and ribbons may appear on exposed surface of slate after it is laid on the roof.

Note—Large quantities of No. 2 Ribbon roofing slates are used for damp courses or dampproofing of masonry construction.

Chapman and Hard Vein Quarries, Pennsylvania

There are two grades of Commercial Standard slate from Pennsylvania Chapman and Hard Vein quarries, viz., No. 1 and No. 2, both having one or more hard veins running across the slate.

Number 1

Surface Reasonably smooth straight cleavage full length of slate both front and back. The maximum bend shall not exceed ¼″ in lengths up to 16″, nor exceed ⅜″ in lengths from 16″ to 24″.

Texture Shall be free from knots or knurls that in any way interfere with the safe conveyance or the laying of the slate on the roof.

Corners Reasonably full corners on exposed ends. No broken corners on covered ends that would sacrifice nailing strength, or the laying of a water tight slate roof.

Thickness Approximately $\frac{3}{16}$″.

Weight 650 lbs. to 700 lbs. per square.

Number 2

Surface Same quality as No. 1 but a rougher surface. The veins showing more prominent. Not quite as uniform in thickness as No. 1.

Texture

Corners Reasonably full corners on exposed ends with no broken corners on covered ends that would sacrifice nailing strength or the laying of a water tight slate roof.

Weight 670 lbs. to 725 lbs. per square.

Vermont and New York

There is only one grade of Commercial Standard Roofing Slates from Vermont and New York.

Surface Reasonably smooth straight cleavage full length of slate both front and back. The maximum bend shall not exceed ¼″ in lengths up to 16″, nor exceed ⅜″ in lengths from 16″ to 24″.

Texture Shall be free from knots or knurls that in any way interfere with the safe conveyance or the laying of the slate on the roof.

Corners Reasonably full corners on exposed ends. No broken corner on covered ends allowed that would sacrifice nailing strength, or the laying of a water tight slate roof.

Weight 650 lbs. to 800 lbs. depending on quarry and color.

Thickness Approximately $\frac{3}{16}$″.

Note—"Smooth" does not mean an absolutely even face and back, because the natural characteristics of the rock render this impossible. The word "smooth" is therefore used relatively only. The rougher slates are selected for use on other than the standard type of roof.

Monson District, Maine

There is only one grade of Roofing Slate from the Monson Maine district.

Surface A smooth slate with grain running the length of slate.

Corners Reasonably full corners on exposed ends.

Weight Approximately 725 pounds per square.

Thickness $\frac{3}{16}$″ standard thickness.

Peach Bottom District

There is only one grade of Commercial Standard Roofing slate from the Peach Bottom district.

Surface Reasonably smooth, straight cleavage out of wind.

Corners	Reasonably full corners on exposed ends. Some corners are allowed off on ends under cover, but not enough to damage its service.
Weight	About 700 lbs. to 750 lbs. per square.
Thickness	Approximately $\frac{3}{16}$″.

Virginia

There is only one grade of Commercial Standard Roofing Slate from Esmont district and other Virginia quarries.

Surface	Exceptionally smooth straight cleavage full length of slate both front and back.
Corners	Reasonably full on exposed ends. No broken corners on covered ends that would sacrifice nailing strength or the laying of a water tight slate roof.
Weight	Practically uniform about 700 lbs. per square.
Thickness	Approximately $\frac{3}{16}$″.

Note—The rougher slates are selected for use on other than the standard type of slate roof.

Quantity

In the United States slates are sold by the "square." A "square" of roofing slate is defined by the U. S. Department of Commerce, Bureau of Standards, in Simplified Practice Recommendation No. 14, as follows:

> "A Square of Roofing Slate—A square of roofing slate means a sufficient number of slate shingles of any size to cover 100 square feet of plain roofing surface, when laid with approved or customary standard lap of 3 inches. Slates for surfacing flat roofs are usually laid tile fashion, without lap, in which case a square of slate would cover an area greater than 100 square feet."

The quantity per square varies from 686 pieces for the 10″ x 6″ size to 98 for the 24″ x 14″ size, which includes the allowance for a 3″ head lap.

It should be noted that for roofs of comparatively little slope where a 4″ lap is required, an additional quantity must be provided. For steep roofs or siding, where a lap of 2″ is sufficient fewer slates will be necessary. Slate, however, is always sold on the basis of quantity required for a lap of 3″ even for flat roofs.

The data under the next three headings is quoted from Simplified Practice Recommendation No. 14.

ROOFING SLATE

In accordance with the unanimous action on January 23, 1924, in New York, N. Y., of the joint conference of representatives of manufacturers, distributors and users of slate for roofing purposes, the United States Department of Commerce, through the Bureau of Standards, recommends that recognized sizes and nomenclature be reduced to those shown below.

TABLE 1.—Dimensions of Slate Shingles for Sloping Roofs; Minimum to a Square

[Each size split[1] to thickness of $\frac{3}{16}$, ¼, ⅜, ½, ¾, 1, 1¼, 1½, 1¾ and 2 inches.[2]]

Face dimensions[3] in inches	Minimum number to square (3″ lap)	Face dimensions[3] in inches	Minimum number to square (3″ lap)
10 by 6	686	16 by 10	221
10 by 7	588	16 by 12	185
10 by 8	515	18 by 9	213
12 by 6	533	18 by 10	192
12 by 7	457	18 by 11	175
12 by 8	400	18 by 12	160
12 by 9	355	20 by 10	169
12 by 10	320	20 by 11	154
14 by 7	374	20 by 12	141
14 by 8	327	20 by 14	121
14 by 9	290	22 by 11	138
14 by 10	261	22 by 12	126
14 by 12	218	22 by 14	109
16 by 8	277	24 by 12	115
16 by 9	246	24 by 14	98

[1]The art of splitting slate blocks consists in progressively reducing resultant halves, until the desired roofing slate thickness has been reached or approximated. This hand-wrought characteristic appeals to architects and owners. It is not a simple matter to precisely control the splitting of this natural rock, nor can a uniformity of thickness throughout be assured. The recommended range of thicknesses to be aimed at by operative splitters will meet all normal requirements, and will insure the maximum of economy in the utilization of the many sizes of quarried blocks.

[2]It is customary to regard a thickness falling between two standard thicknesses as a "special," and it is the practice to base the price of the "special" upon the greater of the two standard thicknesses.

[3]For thicknesses one-half inch and more, it is not generally considered practicable to use lengths that are less than 16 inches, although for roofs of special treatment it may be done in small quantities. *(But on large projects it will be found more economical to allow 16-, 18- and even 20-inch lengths than to specify only the 16-inch minimum.)* In carrying out a desired design on special roofs, it is sometimes necessary to make shingles longer than 24 inches, in which case the thicker slates are used.

TABLE 2.—Dimensions of Slate Shingles for Flat Roofs

[Each size split to following thicknesses: For ordinary service, $\frac{3}{16}$ inch; for promenade and extraordinary service, ¼ inch and ⅜ inch]

Face Dimensions, in Inches		
6 by 6	10 by 6	12 by 6
6 by 8	10 by 7	12 by 7
6 by 9	10 by 8	12 by 8

SIZES OF SLATE FOR MISCELLANEOUS PURPOSES

It is recommended that smaller slate, such as 12 or 14 inch lengths, be used in covering pents, porch and dormer roofs and sides, and garage or other low-roofed buildings. This practice is also recommended even in situations where the main roof is of larger slate.

Owing to the fact that certain sizes of slate may be more available than the size called for in the specification or order, it is recommended that architects, builders, engineers, and contractors provide for alternate selection on usual slate roofing installations. In this connection, particular attention is also called to the increasing use of random widths of the desired lengths, and to the fact that architects are adopting this practice. While slate is plentiful, such practices will bring about the elimination of waste of an important natural resource, and will obviate the necessity of waiting for specified sizes while an accumulated finished stock of other usable sizes is available and accessible.

DIMENSION NOMENCLATURE

COMMERCIAL STANDARD THICKNESS (that is, average or basic.—The terms "3/16-inch slate," "full 3/16-inch slate," or "not less than 3/16-inch slate" indicate a desire for a hand-picked selection, regardless of the added labor and cost. "Commercial standard" is the quarry run of production, and shows tolerable variations above or below 3/16 inch. For the thicker slates, however, reasonable plus tolerances only are permissible; thus a ¼-inch slate must be a full ¼ inch or thicker.

Weight

A square of slate on the roof, *i.e.*, enough slate to cover 100 square feet of roof surface with a standard 3″ lap, will vary from 650 to 8,000 pounds for thicknesses from the commercial standard 3/16″ to 2″.

The commercial standard thickness (approximately 3/16″) will weigh from 650 to 750 pounds per square. For estimating the dead load on the roof construction, it may be taken at a maximum of 800 pounds per square or 8 pounds per square foot to include slate, felt and nails.

Slate used on flat roofs are laid without lap, the ends and sides being butted fairly close together. The exposure on a flat roof is thus the size of the slate used and the weight of slate required to cover a "square" of roof surface is correspondingly less.

The weight of slate varies with the size of the slate, color and quarry and even sometimes in the same quarry. This variation may be from 10% above to 15% below the weights given in the following table of average weights of slate of different thicknesses for both sloping and flat roofs.

Slate Thickness in Inches	Average Weight of Slate per Square (100 sq. ft.)	
	Sloping Roof Allowing for 3″ Lap	Flat Roof Without Lap
Standard 3/16	700	240
Selected full 3/16	750	250
¼	1,000	335
⅜	1,500	500
½	2,000	675
¾	3,000	1,000
1	4,000	1,330
1¼	5,000	1,670
1½	6,000	2,000
1¾	7,000	
2	8,000	

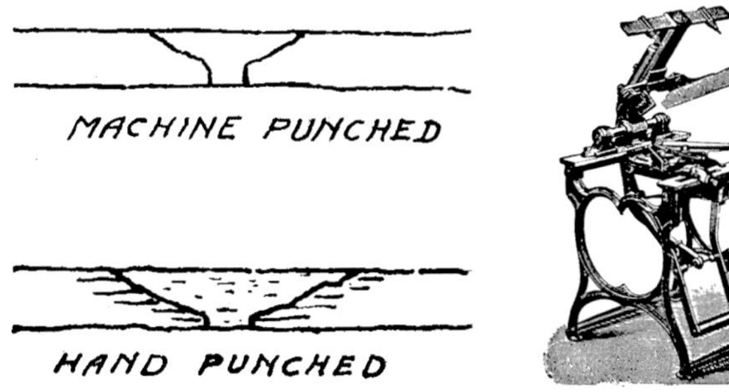

Figure 7

Nail Holes

No slate should have less than two nail holes. The standard practice is to machine punch two holes in all architectural roofing slate ¼″ and thicker at the quarry and also in commercial standard slate when so ordered. Four holes should be used for slates ¾″ or more in thickness when more than 20″ in length. Holes are punched from one-quarter to one-third the length of the slate from the upper end, and 1¼″ to 2″ from the edge. Where four holes are used, it is customary to locate the two additional holes about 2″ above the regular holes.

Some architects in the past specified that all nail holes should be drilled and countersunk. On normal thickness slates, no method of drilling has been developed which will produce the same clean hole as by machine punching, because the small thickness of stone at the cutting point is insufficient to absorb or dissipate the drilling force. Hence the industry, through the Association, in 1923 adopted the machine punching of nail holes as standard practice.

That the results are excellent is demonstrated by the fact that, when given two slates with holes punched by either method, architects have usually selected the machine punched as the better and have even called them drilled and countersunk.

Machine punching is preferable to hand punching. The term hand punching usually refers to the use of the double head slaters hammer having one head in the form of a long prong. Machine punching may be done either at the quarry or on the job. Hand punching of holes in fitting hips, etc., is, of course, necessary.

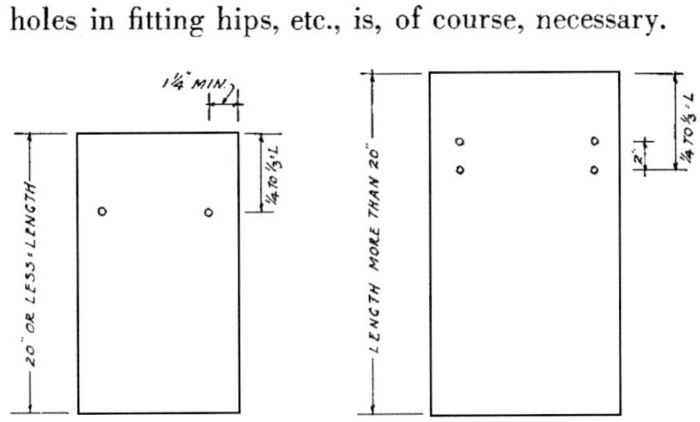

Figure 8. Location of Nail Holes

Laying Slate

IN the laying of any roofing material workmanship is as essential as the proper selection of the material. The more enduring the material the more important this factor becomes. Slate, the most lasting roofing material known, should be laid by roofers of experience and training. It is a mistake to assume that those without such experience are qualified to properly lay slate. For

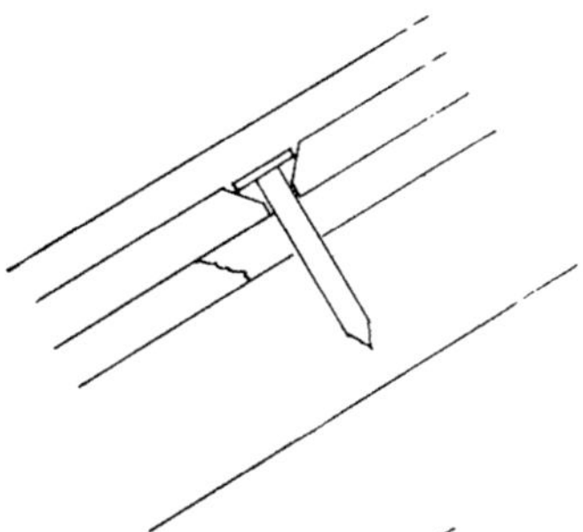

Figure 9.
Properly Nailed

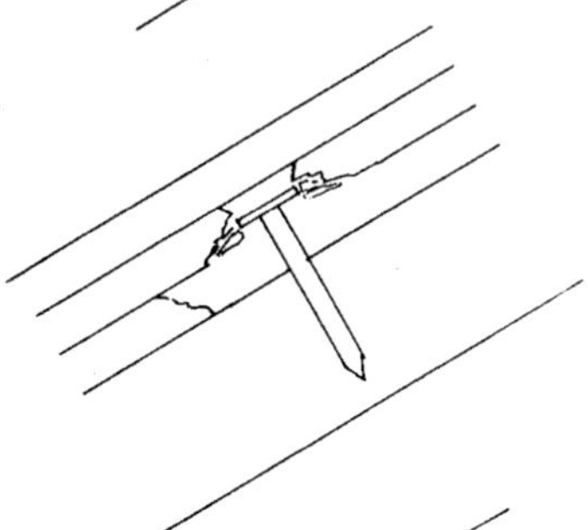

Figure 10.
Nail Driven Too Far

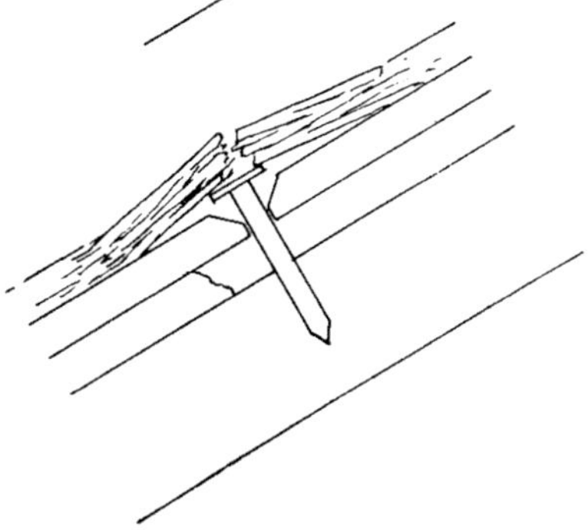

Figure 11.
Nail Not Driven Far Enough

instance the nailing of wooden shingles and slates are entirely different. The heads of slating nails should just touch the slate and should not be driven "home" or draw the slate, but left with the heads just clearing the slate so that the slate hangs on the nail. The opposite is true of wooden shingles and a man used to laying this material will invariably handle slate in the same way. As a consequence the slate, held too rigidly in place, is shattered around the nail hole or the head of the nail crushed and eventually the slate may "ride" up over the nail and be blown off in a heavy wind. The blame is

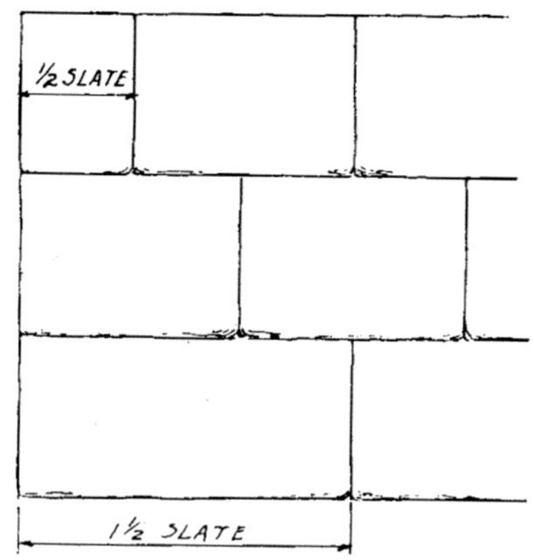

Figure 12. Starting Slate

placed on the material whereas the real reason can be traced to the method of nailing. All nails should penetrate the sheathing and not the joints between boards. This is especially important near the ridge of the roof.

It would seem almost unnecessary to mention the fact that there should be no through joints from the roof surface to the felt. The joints in each course should be well broken with those below. Where this simple precaution is neglected it is possible that water may find its way through the joints, eventually cause the felt to disintegrate and leaks develop. Where random widths are used the overlapping slate should be jointed as near the center of the under slate as possible and not less than

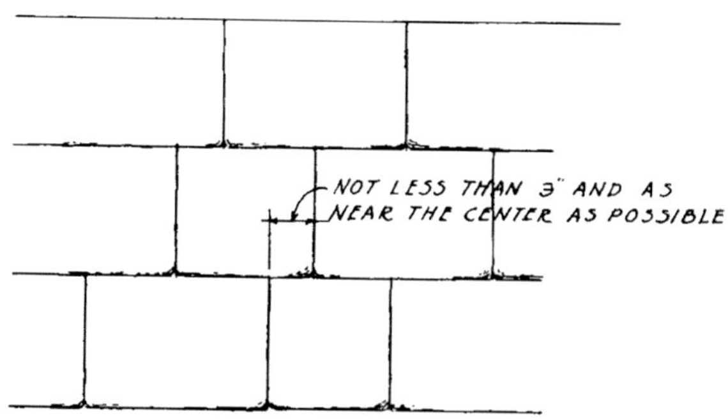

Figure 13. Proper Jointing

3″ from any under joint. Where all slates are of one width, this is automatically taken care of by starting every other course with a half slate or, where available and practicable, a slate one and one-half times the width of the others.

With but few exceptions, the standard 3″ lap should be insisted upon. The "standard 3″ lap" or "3″ headlap" means the lap of the slate over the *second* course below, see Figure 15. The small saving in slate through reducing the lap will not compensate for the risk entailed of leakage due to the lessened amount of material over which water might be blown.

A practice prevalent among many roofers is that of driving the slater's stake into the roof boards. To avoid damaging the roofing felt, a plank should be used for this purpose or the stake driven into the scaffold only. Slaters occasionally use a metal strap as a support for the scaffold brackets. This practice should be discouraged when these are cut off and a part left on the roof. They will rust in time and stain the slate in a most unsightly manner.

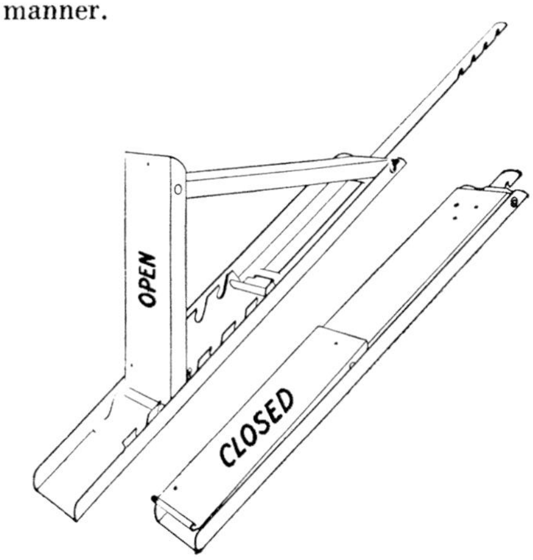

Figure 14. A Satisfactory Scaffold Bracket

The foregoing applies to slating in general. The forming of slate hips, ridges, valleys, eaves and gables require a description peculiar to themselves.

Exposure

The "exposure" of a slate is the portion not covered by the next course of slate above and is thus the length of the unit exposed to the weather. The standard lap of the alternate courses used on sloping roofs is 3″ and is the basis upon which all roofing slate is sold and the quantity computed. The proper exposure to use is then obtained by deducting 3″ from the length of the slate used and dividing by two. For instance, the exposure for a 24″ slate is 24″ minus 3″ = 21″ ÷ 2 = 10½″ exposure.

The following table will be found of use in readily obtaining the proper exposure.

Exposure in Inches for Sloping Roofs

Length of Slate in Inches	Slope 8″ to 20″ per foot, 3″ lap
24	10½
22	9½
20	8½
18	7½
16	6½
14	5½
12	4½
10	3½

Sloping roofs having a rise of 8″ to 20″ per foot of horizontal run should be laid with the 3″ lap. Buildings located in the southernmost parts of the country or on the Pacific slope may however be safely roofed with a lap of 2″ providing a high standard of workmanship is otherwise maintained. For steeper roofs, such as the Mansard and others nearly vertical in plane a 2″ lap will usually be found sufficient. In some sections of the country it is customary to increase the lap to 4″ when the slope is from 4″ to 8″ per foot, while in other parts the 3″ lap is considered entirely adequate. Flat roof construction should be used for slopes less than 4″ per foot. For vertical walls or siding use 2″ lap.

Ridges

There are two common methods of finishing the ridge of the roof. These are usually known as the "Saddle Ridge" and "Comb Ridge" but each may have other names and certain variations in laying according to local practice. In Figure 16 are illustrated two types of saddle ridge which are known respectively as the "Saddle Ridge" and the "Strip Saddle Ridge." In the first of these, the "Saddle Ridge," the regular roofing slates are extended to the ridge so that pieces of slate on the opposite sides of the roof butt flush. On top of the last regular course of roofing slate at the ridge is laid another course of slate called the "Combing Slate" and the pieces on the opposite sides of roof butted flush. The combing slate is usually laid with the grain horizontal and should be of such width that the exposure or gauge

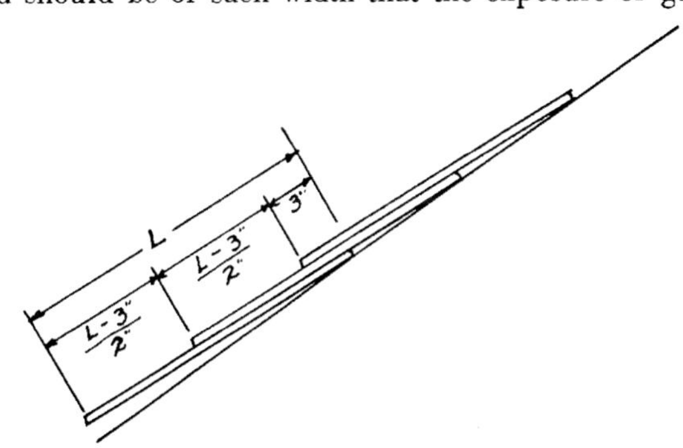

Figure 15. Lap and Exposure

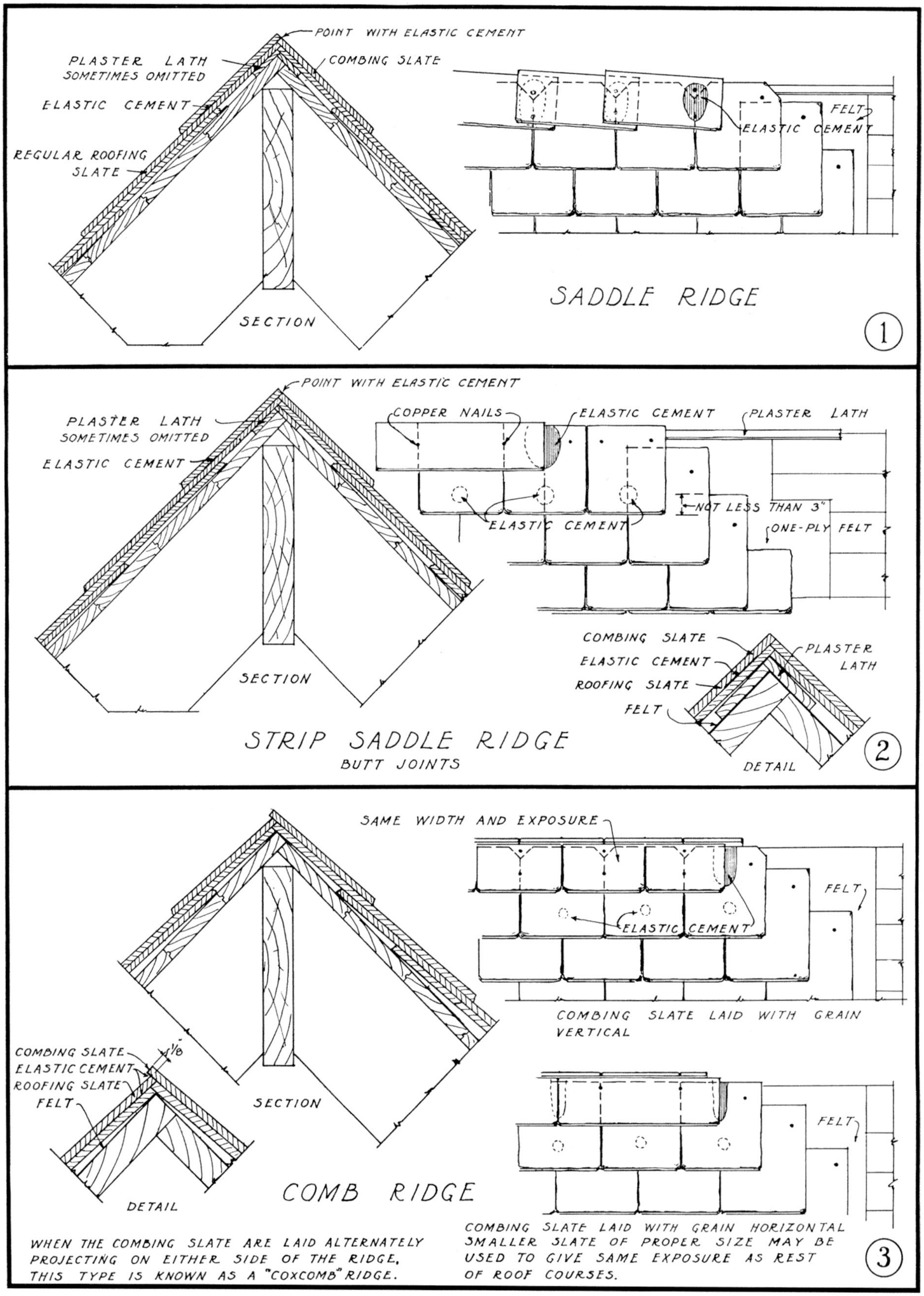

FIGURE 16. STANDARD DETAILS SLATE RIDGES

of the roof is maintained approximately uniform. For example if 20″ x 12″ slates are used on the roof with an 8½″ exposure, 12″ x 8″ slates laid horizontal could be used on the ridge. It will be noted in Figure 16 that the combing slates overlap and break joints with the underneath slate. In this way all the nails in the combing slate are covered by the succeeding slates except the nails in the last or finishing slate on the ridge and these nails should be covered with elastic cement. In Figure 16 it will be seen that only two nails are used in each slate. For this reason the end of the slate which is not nailed should be held in place by elastic cement. The joints on top of the ridge formed by the butted edges of the combing slates should be filled in with elastic cement when subject to heavy rainfall. Some roofers do not use

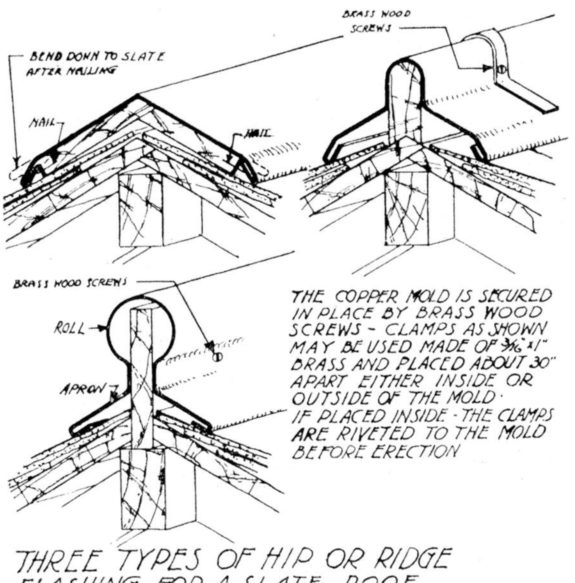

Figure 17

any elastic cement on either the hips or ridges, but this practice is not recommended except under favorable climatic conditions. The nails should be arranged to go between the joints of the slate immediately below. In some parts of the country it is customary to clip the upper edges of the last roof slate as shown in the drawing, Figure 16, Plate 1. Many architects prefer to keep the grain of the slate vertical, using the same type of ridge with top or combing slate of the same width as the regular roofing slate and the length the same as their exposure. In such cases the starting slate could be a "slate and a half" in width rather than a "half slate."

Another type of saddle ridge is that known as the "Strip Saddle Ridge." This ridge is laid in a similar manner to the above except that the combing slates do not overlap but butt flush and each combing slate has

Figure 18. Ridge and Closed Valley

four nails. The combing slate may be the same width as the regular roofing slate or narrower as the designer may wish. The nails should be covered with elastic cement and the edges of the combing slates set in elastic cement as shown in Figure 16.

The "Combing Ridge" is laid in the same manner as the "Saddle Ridge" except that the combing slate of the north or east side extends beyond the ridge line as shown in the detail, Figure 16. This extension should not be more than 1″. This type of ridge may be laid with the combing slate having a grain vertical or horizontal. In either case, the edge of the slate should be set in elastic cement, as shown, and the nails covered with elastic cement. If the top or combing course is projected 1/16″ to ⅛″ above the under top courses, it will make a much better finish and will be more easily filled with elastic cement.

A variation of this type of ridge is known as the "Cox-

Figure 19. Fantail Hip

comb Ridge" in which the combing slate alternately project on either side of the ridge.

It will be noted that in Figure 16 the top courses of the regular roofing slate have the edges set in elastic cement. This is done to avoid their lifting under action of the wind and should in no sense be construed as being necessary from the standpoint of weather-proofness.

Hips

There are several methods of forming hips on slate roofs, some of which are illustrated in Figure 22. These are the ones most common in all sections of the country although they may be known by other names than those given herein.

Figure 20. Open Valley

The "Saddle Hip" may be formed by placing on the sheathing forming the hip one or two plaster lath or a 3½" cant strip and running the roofing slate up to this strip. On top of the cant strip and the slate are laid the hip slates which are usually the same width as the exposure of the slates on the roof, although they may vary in width on different classes of work.

It will be noted in the detail Figure 22, Plate 1, that the four nails used to fasten the hip slate to the roof are driven into the cant strip and do not go between the joints of the slate. The heads of these nails are then covered with elastic cement and the lower part of the next slate bedded therein as shown. Elastic cement is also placed on the joint between the roofing slate and the plaster lath and on the peak of the hip before the hip slates are laid. A variation of the "Saddle Hip" is known as the "Strip Saddle Hip" which is used on less expensive work and may be formed of narrower slates laid with butt joints which do not necessarily line up with the course of the slate on the roof.

Another type of hip is that known as the "Mitred Hip." In forming this type of hip the slates forming the roof courses and the hip are all in one plane as is shown in Figure 22, Plate 2. The hip slates should be cut accurately to form tight joints and the joint should be filled in with elastic cement. The nail holes should be so placed as to come under the succeeding hip slate.

A variation of this type of hip is that known as the "Fantail," shown on Figure 22, which is laid in the same manner as the "Mitred Hip" but which has the bottom edge of the hip slate cut at an angle to form a fantail.

Another very popular type of hip is that known as the Boston Hip. In this type of hip the slates are woven in with the regular courses of the roofing slates, as shown in the detail. The nails are then covered with elastic cement and the lower part of the succeeding slate bedded therein.

It is sometimes recommended that metal or slip flashings be woven in with each course of "Mitred Hips" but this is usually unnecessary if proper care and workmanship are exercised in cutting, fitting and bedding the hip slates. There are some roofers who do not use elastic cement on the hip slates and secure satisfactory results.

Valleys

Of the two methods of forming the valleys the first, and without doubt the more satisfactory, is the open valley. The second, known as the closed valley, is considered by many to be the more pleasing in appearance and is much used on high-grade work. Variations of the closed valley, frequently used in connection with the Graduated or Textural roofs, are the "round" valley and the "canoe" valley.

OPEN VALLEY

The open valley is formed by laying strips of sheet metal in the valley angle and lapping the slate over it on either side, leaving a space between the slate edges to act as a channel for water running down the valley angle. The width of the valley, or the amount of space between the slate edges should increase uniformly toward the bottom. The amount of this increase, or taper, has been determined as 1 inch in 8 feet. For example, in a valley 16 feet long, the distance between slates will be 2 inches greater at the bottom than at the top, as the width in-

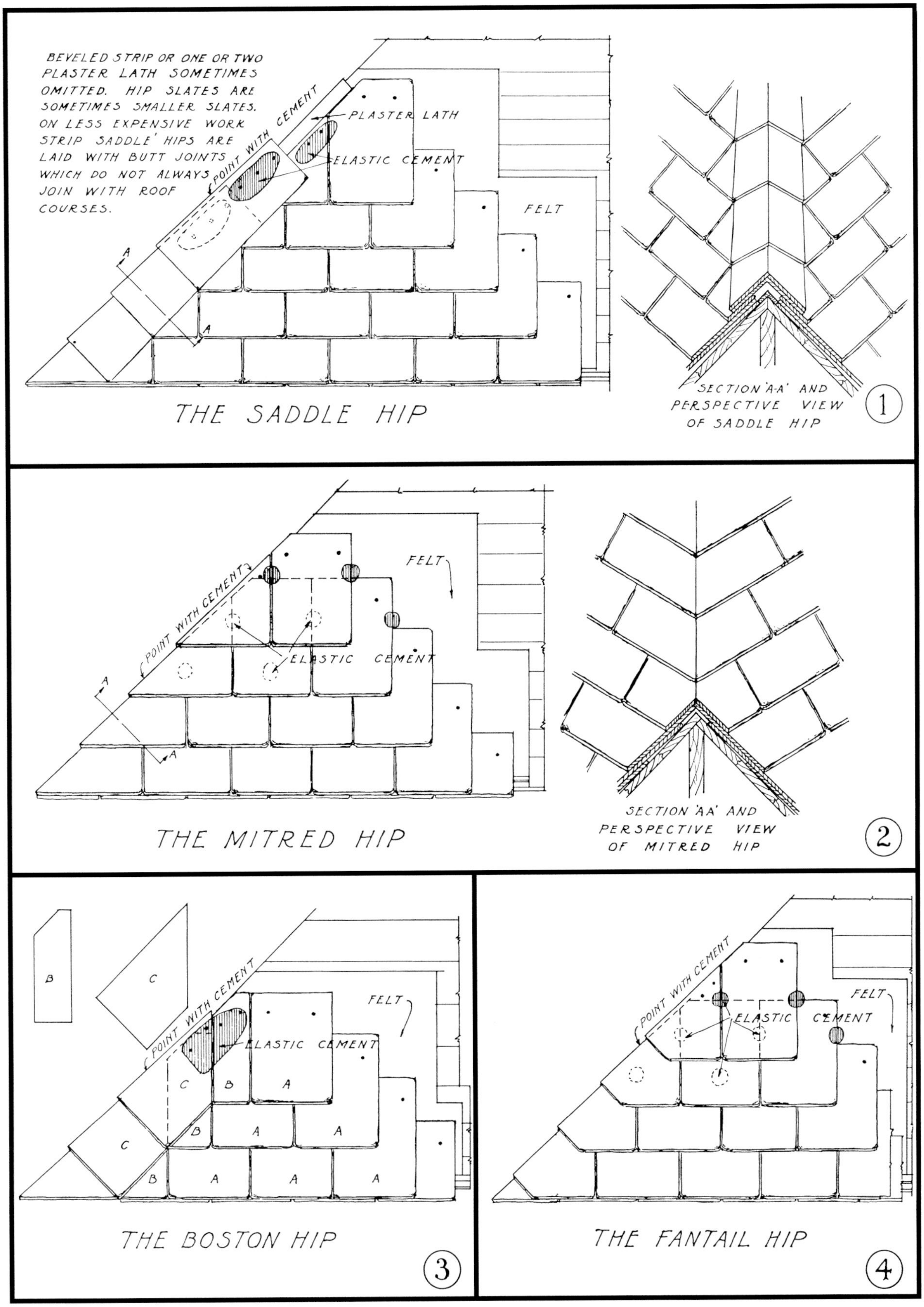

FIGURE 22. STANDARD DETAILS SLATE HIPS

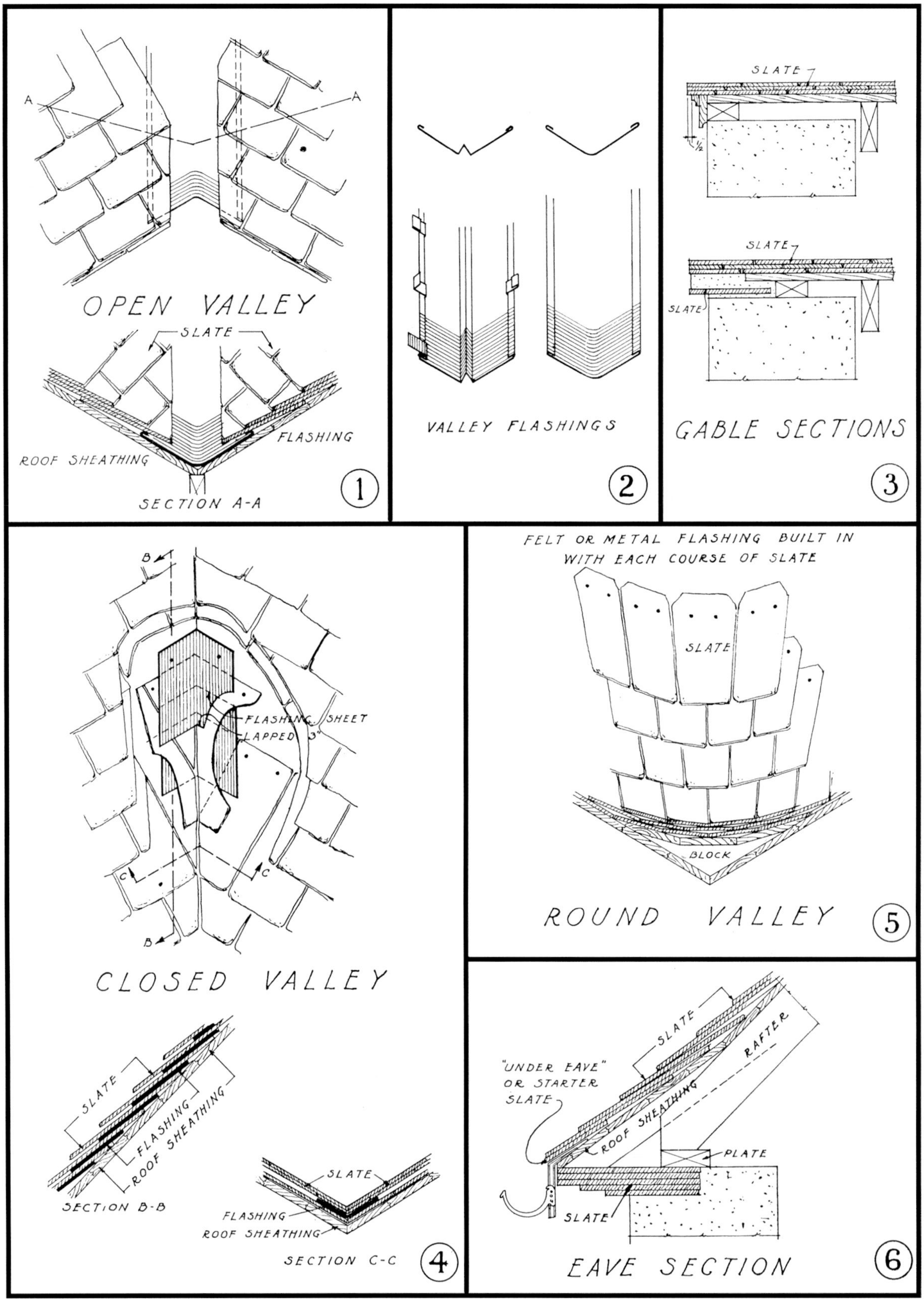

FIGURE 23. STANDARD DETAILS SLATE VALLEYS, EAVES AND GABLES

Figure 21. Closed Valley

creases at the rate of ½″ in 8 feet on each side of the valley. This permits a uniform width of about 2/3 the width of the slate under the slate adjacent to the valley. The difference in width or taper allows the slate to be laid closer to the valley at the upper than the lower end and takes care of the increase in water received. This tapering of the valley also has the very practical effect of allowing any ice which may form to free itself and slide down as it melts.

This increase of width in the valley must be allowed for in placing the flashing strips. Valley flashings are generally laid in pieces up to 8 feet long. The best theoretical manner of taking care of the taper would be, of course, by tapering the sheets. As this involves considerable additional expense in labor and material it is often more practicable to use sheets shorter than 8 feet and increase the width of each sheet an amount sufficient to take care of the taper. This increase in width amounts to 1/8 of an inch per foot. The increase in the widths of succeeding sheets of various lengths necessary to take care of this taper is shown in table No. 3. Figure 23 shows the method of installing the sheets.

The slate should start 2″ each side of the valley center at the top and should taper away from the center at the rate of ½″ for every eight lineal feet. The metal flashing should be of sufficient width to extend up under the slate not less than 4″ (preferably 6″ to 8″), and as far as is possible without being punctured by the slating nails.

Where the two roofs forming the valley have considerable difference in slope or the roofs are much different in size and cause a large variation in the volumes of water to be delivered into the valley, the metal should be crimped or made with a standing seam to break the force of the water from the steeper or longer slope and prevent its being driven up under the slate of the opposite side.

TABLE 3

Showing increase of widths of untapered sheets in a tapered valley.

Length of Sheets	Increase in Width	Length of Sheets	Increase in Width
24	$\frac{1}{4}$	60	$\frac{5}{8}$
30	$\frac{5}{16}$	66	$\frac{11}{16}$
36	$\frac{3}{8}$	72	$\frac{3}{4}$
42	$\frac{7}{16}$	84	$\frac{7}{8}$
48	$\frac{1}{2}$	96	1
54	$\frac{9}{16}$		

EXAMPLE: A valley is 19 feet long. The sheet extends 5 inches under the slates and is fastened by cleats. What width of sheets can best be used? As a 4-inch minimum under the slate is necessary two eight- and one three-foot length can be used. Starting at the top the first sheet would be 2+2+5+5+½+½=15 inches wide. If the three foot is used at the top, the first eight foot sheet would be 15⅜ or 15½ inches wide and the second one 16⅜ or 16½ wide.

Condensation forming on the underside of valley flashings, when not free to run off or evaporate, may attack the metal. It is therefore recommended that the felt be omitted under metals if other than copper is used. If felt is used under other metals they should be well painted on the underside.

For inexpensive roofs as provided for in the specification for a Standard Roof the copper for valleys is laid flat without crimps or cleats. For high-grade work the copper sheets should be secured to the roof boards and over the felt with metal cleats from 8″ to 12″ apart.

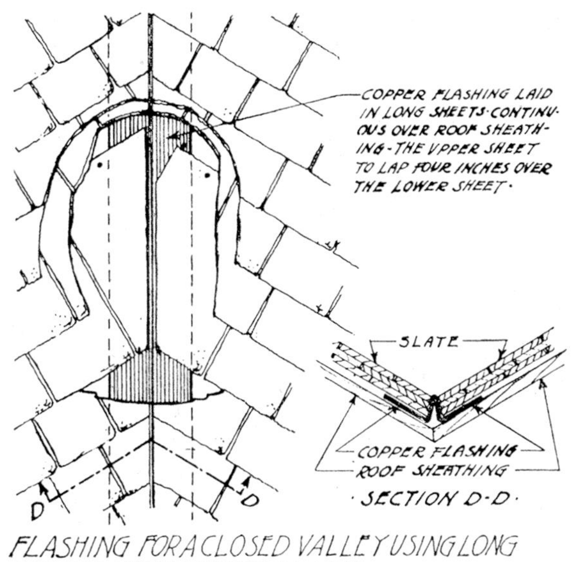

FLASHING FOR A CLOSED VALLEY USING LONG SHEETS UNDERNEATH SLATE

Figure 24. Another Type of Closed Valley

The edge of the sheet is turned over ½″ and the bent end of the cleat hooked under. The cleat is then nailed to the roof boards with two nails and the cleat bent over to cover the nails. A method sometimes used on wide valleys with the best work is to fold the metal 4″ or 5″ from the valley line and 3″ from the cleat fold. This is known as "Fold-over" flashing.

CLOSED VALLEY

The closed valley is formed with the slate worked tight to the valley line and pieces of metal placed under the slate as shown in Figure 23. The size of the sheet to be used is determined by the length of the slate and the slope of the adjoining roofs. Each sheet should extend 2″ above the top of the slate on which it rests so that it may be nailed along the upper edge of the roof sheathing without the nails penetrating the slate. Each sheet should be long enough to lap the sheet below at least 3″, and should always be set back of the butt of the slate above so that it will not be visible. These sheets are separated by a course of slate (Section BB). Each sheet must be wide enough so that the vertical distance from the centre of the valley to a line connecting the upper edges of the sheet will be at least 4″. This dimension depends upon the nailing of the slate which should not penetrate the sheets.

Some roofers form the sheets with a center crimp (Figure 24), thus stiffening them and forming a straight line to which to set the slates, and preventing water from one slope forcing its way above the sheet on the other slope.

Another method of forming a closed valley is shown in Figure 24. The sheets are laid in long pieces directly on the paper or felt covering the roof sheathing before the slate is laid. They may be of any desired length and should lap in the direction of the flow at least 4″. They should be nailed about every 18″ along the outer edge, and care should be taken to avoid penetrating the sheet when nailing the slate.

Figure 25. Round Valley

Figure 26. Eaves and Gables

ROUND VALLEY

The round valley forms a pleasing transition between two intersecting slopes when used in connection with the Graduated or Textural roof. However, if not properly laid out it will produce disaster and mar an otherwise beautiful roof. For this reason its laying should be entrusted only to those fitted by knowledge and experience to do this particular work. It requires the most careful workmanship and experienced knowledge of the problem to secure a job that will be both pleasing in appearance and water-tight.

Primarily the round valley requires a suitable foundation to establish the general contour. The method of building this foundation is described under sheathing and roof boards on page 40.

The valley slates must be at least 4″ longer than the slates used in the corresponding courses of the roof. The sides of the slates must be trimmed to the proper radius and the tops shouldered to make the slates lay flat.

The round valley slates are sometimes bedded in elastic cement. If proper care is used in the trimming and fitting, no flashings should be necessary. Where the workmanship is not dependable, flashings of metal or prepared roofing cut to the proper radius should be used as a precautionary measure. Flashings should always be used wherever ice may form.

The radius of the round valley starts as a maximum at the eaves and gradually diminishes to practically zero at the ridge. For appearance, as well as to facilitate laying the valley slates, the distance across the eaves should not be less than 26″. If the roof condition will not permit this, the "canoe" valley should be used.

The canoe valley is a variation of the round valley and is laid in the same manner except that the radius at the eaves and ridge is practically zero. The radius is gradually increased until it becomes a maximum halfway between the eaves and the ridge.

Eaves and Gables

The under-eave slates should start on a cant strip of suitable thickness, depending upon the thickness of the slate, to enable the second course of slate to be correctly laid. In the case of a cornice, this slate should project about 2″ beyond the cant strip, sheathing or finishing member. The length of the under-eave slates is found by adding 3″ to the exposure being used on the regular slates. Thus, if 16″ slates are used, the exposure is 6½″ and the size of the under-eave slate required is 9½″. Half slate are sometimes used, or roofing slates of the proper width may be laid horizontally. If the first course is ¾″ in thickness, use ⅜″ slates for the under-eave course or ¼″ slates if the starters are ½″, although the under-eave and first course are sometimes made the same thickness.

The first course of slate is laid over the under-eave course with the butts of both courses flush, and the joints broken.

Figure 27. Laying Slate

When changing from a roof of flat slope to one of steeper slope as in the case of a Gambrel roof, the slate of the upper and flatter roof should project 2″ to 2½″ beyond the steeper roof below. A cant strip should also be used upon which to start the slate of the roof of lesser slope the same as at the eaves.

At the gables the slate should overhang the finishing member of the verge board not more than ½″. Where close-clipped gables are used or the construction is such that the gable slates have ample nailing, this dimension may be increased, but the projection ought not to be too great for good appearance. Also there are many interesting ways to lay "Gable end" or "Barge" slates under regular courses along gable ends where shadow effect is desired. Ways and means of using and securing all gable end slates depends on type of construction.

Roofing Felt

It should be emphasized that a standard slate roof can be laid water-tight on open lath without felt, as is often done in the South or on buildings where heat is not required. The thickness of the felt has little relation to the water-tight qualities of a slate roof. The opinion has been held by many that the thicker the felt, the tighter will be the roof. When it is realized that every nail used in fastening the slate is driven through the felt it will be seen how erroneous is this idea. Roofing felt, however, does have three other distinct uses in connection with slate roofs. The felt placed as soon as the roof is sheathed will protect the building, when necessary, until the slates are laid.

It has considerable insulating value in resisting the heat of the summer sun and the transference of heat from the inside during the winter months. Increasing heating costs in the colder localities makes this a factor worth careful consideration. The use of laths over the felt and under the slates to obtain an air space, a method recommended by English roofers, adds much to the insulating value of the felt.

The third value of the felt is to form a cushion for the slate. While not of great moment in the case of commercial standard slates, its value in this respect increases as the thickness and weight of the slate increases. For the commercial standard slate felt weighing 14 pounds per square will be found satisfactory. The same weight is ordinarily used under Textural roofs. Any of the heavier weight felts, as 20, 30 or 40 pounds per square, may be used where the appropriation allows. For Graduated roofs, the 30 pound weight is commonly used when the slates are ¾″ or less in thickness and the 50 pound for slates 1″ or more. It is customary in some localities to place two layers of felt under the slate on a

Graduated roof. The first layer is usually 30 pound felt and the second 14 pound. This provides an extra cushion for the heavy slates. The joints and laps should always be staggered.

The felt should be laid in horizontal layers with the joints lapped toward the eaves and at the ends. A lap of at least 3″ should be used and the edges should be well secured to the surface over which it is laid. A lap of not less than 2″ should be used over the metal lining of valleys and gutters. If metal other than copper is used as a lining the felt should be omitted in the valleys. Extend the felt over all hips and ridges at least 12″ to form a double thickness.

Asphalt saturated rag felt should always be used. The so-called "Slaters' " felt includes many types of materials which cannot be recommended.

Elastic Cement

In the past the extensive use of Elastic Cement has often been recommended because it was supposed to be necessary in order to make a slate roof water-tight. It is now realized that this is erroneous and that it is not only possible but practical to omit Elastic Cement entirely and in some localities experienced roofers have abandoned its use and secured absolutely water-tight roofs. It has even been found that in certain locations on the roof the use of cement may prove to be detrimental rather than helpful. An instance of this is under slates adjacent to open valleys where the cement may dam the water and force it back under the slates instead of permitting it to run out into the valley.

Elastic Cement does have its value, however, and is used under slates forming hips and ridges to help hold securely in place those slates which are usually smaller than the regular roofing sizes and which cannot be so well nailed. Elastic Cement is also used for pointing the peaks of hips and ridges.

The requirements which should govern the selection of Elastic Cement are few but important. It should be water-proof. It should have a high melting point to prevent the slates from slipping under the heat of the sun and a low freezing point to prevent its becoming brittle and cracking in cold weather. It is important that the cement should not dry out and pulverize when exposed to the air. The best grades of cement are of an oily, sticky nature and, considering the small amount ordinarily required for a slate roof, it is economical to use only the best quality.

The Elastic Cement should match as nearly as possible the color of the slate, but it should be noted that cements of certain colors may fade upon exposure, and experience only will show those that are fast and satisfactory.

Slater's Tools

The tools commonly used by the slater are the punch, hammer, ripper and stake. If much punching is done at the yard or on the job by the roofer, a punching machine should be used for punching the nail holes and cutting the slates. It is adjustable to any size or shape, cuts and punches at one operation with a countersunk hole. The hand or mawl punch is forged from fine tool steel hardened and ground about 4½″ long with one end tapered. The butt end is struck with a mawl to punch the nail hole.

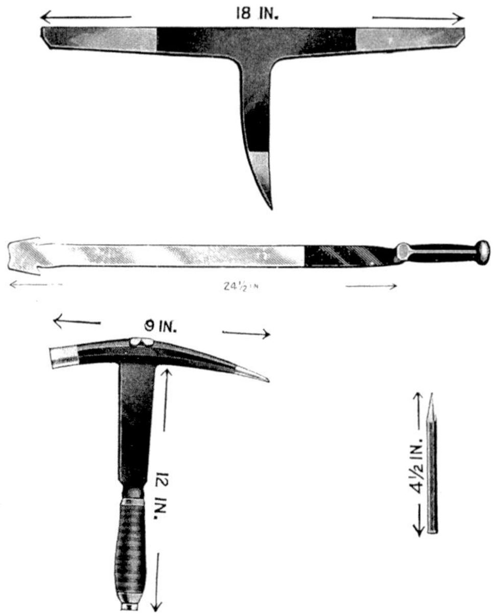

Figure 28. Slater's Tools

Slater's tools are all drop forged. An approved hammer is forged solid, all in one piece, from crucible cast steel, with an unbreakable leather handle to avoid slipping and blistering the hands. One end terminates in a sharp point for punching slate, the other in the hammer head. There is a claw in the center for drawing nails, and on each side of the shank there is a shear edge for cutting slate. The head, point and cutting edges are properly tempered to withstand heavy work. The slater's stock size hammer has a 12″ handle.

The ripper is about 24″ long and is forged from crucible cast steel. It is used for removing broken slate and making repairs. A hook on the end provides a means of cutting and removing the slating nails. The blade is drawn very thin and the hook end correctly tempered for hard wear.

The stake is about 18″ long and T-shaped. The long edge is used as a rest upon which to cut ~~and~~ punch slate or as a straight edge to mark the slate when cutting and fitting around chimneys, hips, valleys, etc. The short arm is tapered and pointed for driving into a plank or scaffold.

These tools, as well as 24″ stakes, shorter rippers, left-hand hammers or special hammers or tools, can be obtained from any slate producer distributing roofing slate or from manufacturers of such tools. The slater's equipment is completed with a nail pouch, tinner's snips, rule and chalk line.

Nails

Slate, being a permanent material, is worthy of care and thought in the proper selection of the various materials used in connection with it, and especially as to the method of securing the slate to the roof construction.

Like any other construction unit, a slate roof can only be as strong and enduring as its weakest part, and the majority of slate roof failures over a period of years may be attributed to the punching of the nail holes, the nailing of the slates, or the nails themselves. As has previously been stated, the art of properly laying and nailing slate is not to be discounted and belongs to men trained especially in the work. The punching and nailing of the slate have been described under the heading, "Laying Slate," on page 13.

Before nails came into extensive use, the slate were held in place by means of wooden pegs driven through the slate and hooked over the roof lath. It is the practice in some localities today to hang the slate to the laths or battens by means of heavy wire hooked through the slate and over the laths. This method is in general use where the slate is laid directly on steel construction. Copper nails of sufficient length to be securely hooked and clinched over the structural angles may also be used. These should have large heads and the shafts be of No. 10 or 11 gauge metal.

Nailing is more extensively used today than other methods for securing the slate, and careful attention should be given the characteristics of nails selected for this purpose. The important considerations involved are shape, size and material.

For all practical purposes, the ordinary diamond point and smooth shaft are sufficient for a slating nail and the needle point is seldom, if ever, necessary or of advantage. The shaft, since it supports a greater weight and must resist a small shearing stress, should be larger than that of the shingle nail. To prevent the slate from being lifted up over the nail after being laid, the diameter of the head should be greater than that of shingle nails.

The temptation to use shingle nails instead of slating nails should be discouraged, for the slight saving in cost on the entire roof cannot approach the cost of repairs which may develop as a result of this practice. Architects and owners should insist that the roofer use nails of proper size and kind of non-ferrous metals.

The much-mooted question of the material of which the nails should be made must remain a matter of opinion and judgment until an impartial investigation shall throw further light upon the subject.

It is hoped that research in this field may be undertaken in the near future and definite results furnished those interested. It is a generally accepted fact that copper is one of the most enduring of metals and that iron and steel, adequately protected from corrosion by a heavy coating of zinc applied by the hot-dipped process, will give reasonable service. Plain or ordinary galvanized nails should not be used for laying slate. Nails having a copper content, such as "yellow metal," or "Muntz Metal" and cut zinc nails are sometimes used. Nails should be carefully selected and be the best grade of a reputable manufacturer. Recently "Cimet," "Everdur" and similar chrome-iron alloy nails and other types particularly suited to resist atmospheric corrosion, have been put on the market. Their cost is higher than copper, yet for certain buildings with excessive or unusual acid fumes under and surrounding the slate roofs, it may prove economical to use such nails. Time and wider use of these newer types will prove whether they are or are not superior to copper. When cost is an item, the "Copperweld" nail, being less expensive than solid copper, is often used and may prove to be the satisfactory method of protecting the steel shaft.

Under ordinary conditions, it will be found satisfactory to use 3d nails for commercial standard slates up to 18 inches in length. Use 4d nails for the longer slates and 6d on the hips and ridges. Thicker slates require longer and heavier gauge nails. The proper size may be determined by adding 1 inch to twice the thickness of the slate. Where the through penetration of the sheathing

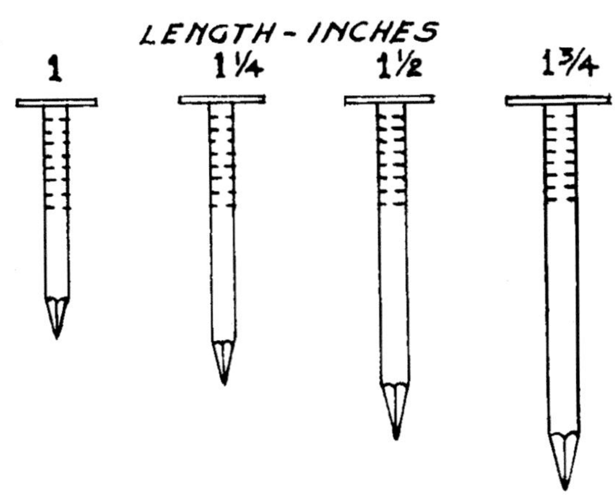

Figure 33. Full Size Nails

will not ordinarily be seen, it provides evidence that nails of sufficient length were used. Thus a ¼″ slate will require a 4d nail which is 1½″ in length. Where the under side of the roof boards is exposed to view, as is sometimes the case of overhanging eaves, a nail of such length as will secure sufficient penetration but not be driven through the sheathing must be selected.

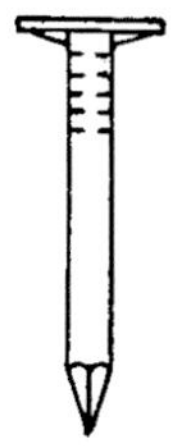

Figure 29 Copper Wire Nail. (Similar to Steel Wire Nails)

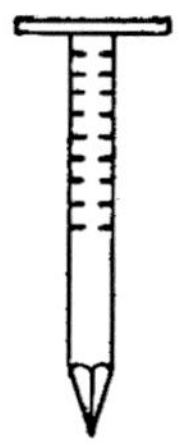

Figure 30 Large Flat-Head Copper Wire Nail. (Slating Nail)

Figure 31 Regular Cut Copper Nails

Figure 32 Large Flat-Head Cut Copper Roofing Nail

[Note difference in head of ordinary wire nail (Figure 29) and large flat-head wire nail (Figure 30)].

Nails suitable for roofing purposes are made in four forms, each of which has its advantages.

The common wire nail (Figure 29) is used generally for nailing flashings, sheathing, and sometimes for shingling. It is not suitable for slate work, for it is of light gauge and small head.

The slating nail (Figure 30), as may be seen, is especially adapted for slating, as it is of heavy gauge and has a wide, flat head. These features make it much more desirable than the common wire nail.

The roofing nail (Figure 31) is not recommended. While the shaft is of proper thickness, its head is too small.

The cut nail (Figure 32) is made from sheets and is of quite different shape than any of those made of wire. The enlargement of the shaft gives it more stiffness than the wire nail has, but there is some danger of splitting the slate if too large a nail is used.

The following tables of slating nail sizes will be found useful in estimating or specifying:

"Copperweld" Slating Nails

Size	Length Inches	Diameter Inches	Nails Lb. (Approx.)
2d	1	.102	386
3d	1¼	.114	211
4d	1½	.128	176
5d	1¾	.128	133
6d	2	.144	87

Copper Wire Slating Nails

Length	Gage	No. per lb.*
⅞	12	303
1	12	270
1¼	10	144
1¼	11	196
1¼	12	231
1½	10	134
1½	12	210
1¾	10	112
2	10	104
2½	8	46

*Add 5% to above for Brass Nails.

Cut Copper Slating Nails

Length	Weight	No. per lb.	Length	Weight
1	2d		2½	8d
1¼	3d	190	2¾	9d
1½	4d	135	3	10d
1¾	5d	100	3¼	12d
2	6d			
2¼	7d			

Cut Brass Nails

Length	Weight	No. per Lb.	
		Slating	Roofing
1⅛		172	
1¼	3d	164	
1⅜		144	
1½	4d	140	216
1¾	5d	108	172
2	6d	88	132
2¼	7d	80	
2½	8d	64	112
2¾	9d	52	
3	10d	48	75
3½	16d		66

Cut Yellow-Metal Slating Nails
(60-40 mixture)

Length	Weight	No. per Lb.
1¼	3d	154
1½	4d	140

An Up-to-date Photo of a Crude Old Slate Date. A more precise design worked out in two colors of slate may provide permanent roof markings for aviators.

FLASHINGS

THE materials commonly used for flashings are copper, tin, lead, galvanized iron and zinc. Flashing materials should be selected with the same care as the other materials entering into the roof construction. The first consideration should be durability. When used with a material as permanent as slate, it is, of course, poor economy to use any but the most permanent and non-ferrous metals and the best workmanship.

Copper

Copper is peculiarly suitable for all flashings (including valleys) as it is easily worked and shaped and adjusts itself to temperature stresses. Copper sheet is made in a wide range of weights and thicknesses and in what is known as soft or "Roofing Temper" copper and hard or "Cornice Temper" copper. Soft copper only should be used for flashings. It is generally defined by the ounce weight per square foot; that is, "16-ounce copper" means copper sheet weighing 16 ounces or one pound per square foot. All copper sheets used should be rolled from copper conforming to the Standard Specifications of the American Society for Testing Materials.

Sixteen-ounce copper sheet is the minimum weight that should be used for flashings of any kind. Many architects will specify nothing lighter than 18-ounce material, and with the best work, where heavy sheets are used, 20-ounce metal is recommended as better practice.

Flashings lighter than 16-ounce are undesirable. All rain water carries with it off the roof dust and grit which have some erosive effect on the metal. It is well to use metal thick enough to do the work of carrying away the water for a period of time at least as long as the life of the building. Sixteen-ounce copper will meet such conditions; 14-ounce is too light.

The following table will be found of value in specifying the weight of sheet copper. Copper for flashings should always be specified by weight and never by gauge.

Weight Oz. per Sq. Ft.	Thickness Inches	Nearest Gauge		
		B. & S.	Stubbs	U. S.
24	.0324	20	21	21
20	.0270	21	22	23
18	.0243	22	24	24
16	.0216	23	25	25

Copper sheets may be obtained in widths from 6″ to 108″ and in lengths from 6″ to 200″. Roll or strip copper is made in widths from 2″ to 20″ and in rolls of 75′ or strips from 6′ to 10′ long. Sheets from 24″ to 42″ wide and from 60″ to 96″ long are in general use in the sheet metal trade and are carried in stock. All specifications and details should be drawn with these sizes in mind.

The edges of all copper flashings to be soldered must be tinned 1½″ on both sides and the seams thoroughly sweated with solder. Proper care in making the seams tight is of utmost importance. Use pieces up to 96″ in length, and except on steep slopes (15° or over), lock and solder all base flashings. Cap or counter flashings need not be soldered. The joint is made by lapping the sheets in the direction of the flow. When there is any likelihood of deep, wet snow packing in or of wind lifting the cap flashings, the joints should be soldered, using either locked or lapped seams.

All exposed edges of flashings—such as the bottom edge of cap flashings—should have a ½″ fold back under for stiffness against wind action. This is a practice that should be axiomatic with good flashing.

Copper requires no painting or other treatment unless it is desired to hasten the development of the natural green patina. In this case it is absolutely necessary that all the grease and oil used in the manufacturing process be removed from the copper. A strong soda solution (4 to 6 ounces per gallon of hot water) will do this. A uniform finish will not be obtained unless the copper is thoroughly cleaned.

Copper can be painted provided the surface be thoroughly cleaned and roughened. This can be done by washing the copper with a solution of 4 ounces of copper sulphate in ½ gallon of lukewarm water in a glass or earthen vessel, to which has been added ⅛ ounce of nitric acid. Before painting, the surface must be carefully washed with clean water to remove the last trace of the solution. For additional data regarding coloring copper, refer to Part 3 of "Copper Flashings," of the Copper and Brass Research Association, or write that association direct at 25 Broadway, New York City.

Tin

Tin used for flashings is more properly known as "Terne-Plate." The base is of iron or steel and the coating a mixture of lead and tin put on the sheet by the hand-dipped process or Patent-roller process. The base metal recommended for flashings is the IX thickness, weighing about 62½ pounds per 100 square feet, or from 250 to 260 pounds per box of 112 sheets size 20″ x 28″. The lighter weight, or IC thickness, may be used but is seldom satisfactory. The weight of the base gives body to the metal, but its enduring qualities depend mainly upon the weight and thoroughness of the surface coating. Heavy coated tin will weigh from 290 to 300 pounds, or represent about 40 pounds of coating. While tin is sold in sheets of various sizes from 10″ x 14″ to 20″ x 28″, the two sizes in more common use are 14″ x 20″ and 20″ x 28″. IX tin is approximately 28 gauge (U. S. standard) and averages about 12 ounces per square foot. IC tin is 30 gauge and weighs about 10 ounces per square foot.

All joints should be securely locked and joints and seams thoroughly sweated with solder.

Tin should always be thoroughly painted one coat on the under side before laying, and all grease and dirt cleaned off, and then painted one coat on the top side after laying. Particular attention should be paid to this under-side coat and a heavy coat applied, as this protects the flashings from the effects of condensation. Metallic brown, Venetian red, red oxide or red lead may be used, mixed with pure linseed oil. No turpentine or dryer should be added. A second coat should be applied to the surface two weeks after the first coat. One or two additional coats may then be applied to obtain the desired surface color. Tin flashings should be gone over, repaired and repainted about every three years. Where dirt or leaves lodge and are retained on the flashings, it is advisable to remove any such accumulations and repaint at yearly intervals. Proper maintenance will add immeasurably to their life.

Lead

The use of lead for building purposes is not new, and it is frequently very desirable for flashings. It is unaffected by ordinary atmospheric conditions and its softness, pliability and malleability make it especially valuable in places where other materials cannot be easily introduced. Until the introduction of hardlead, the only lead available was soft lead, which, while possessing many excellent qualities, was impractical for flashing because of its low physical strength. Hardlead has a much greater tensile strength which permits its use in comparatively thin sheets. Lead is protected by nature through oxidation of the surface upon exposure, and requires no further treatment.

Hardlead is rolled in sheets 24″, 30″ and 36″ wide and 96″ long, weighing 2½, 3, 4, 6 and 8 pounds per square foot. With respect to the proper weight of hardlead sheets to be used, one manufacturer states that this depends upon the purpose, and adds:

"For gutter linings, cornice coverings, base flashings and roofing purposes generally, the 3-pound sheet is recommended, and for cap flashings and batten roofs where the battens are spaced 18″ or less on centers, the 2½-pound sheet may be used."

The lead should be so installed that it can expand and contract, and nailing directly through the sheet should never be permitted. The sheets should be fastened by means of cleats. These cleats should be made of 16-ounce soft rolled copper or 3-pound hardlead, fastened to woodwork with two hard copper wire nails and to masonry with brass screws and lead shields. The cleats should be spaced about 8″ on centers, but on steep roofs continuous cleats for the horizontal joints are recommended.

Where the edge of the metal is fastened by means of a reglet, there should be a continuous cleat of 3-pound hardlead caulked into the reglet and the sheet should be locked to the reglet. Never caulk the sheet into the reglet.

Where the edge of the metal is unfastened, such as cap flashings and similar conditions where a lapped joint is provided, the free edge of the metal should be hemmed about ½″.

All nails should be hard copper wire flat-head nails not less than ¾″ long. All screws should be of brass and all shields of lead; iron or steel nails and screws, coated or uncoated, should not be used. The sheets should be joined together by means of locked seams. Lapped and soldered seams are not recommended. Wooden tools should always be used in working and beating the material into place.

Zinc

Zinc for roofing and general sheet metal work has been used in Europe for more than a century, where its permanence and freedom from repairs have been thoroughly proven. Its use in this country, while of more recent years, has shown similar results.

Zinc is a metal, not an alloy of other metals, which is extremely resistant to the corrosive action of the elements. It rapidly acquires a protective coating (a basic carbonate of zinc), which will continue to form as long as there is any raw zinc exposed. This protective coating gives the metal a light battleship grey color which will deepen with age and approach the color of grey slate. Zinc does not need paint as a protection, but paint can be readily used if other than the natural color of

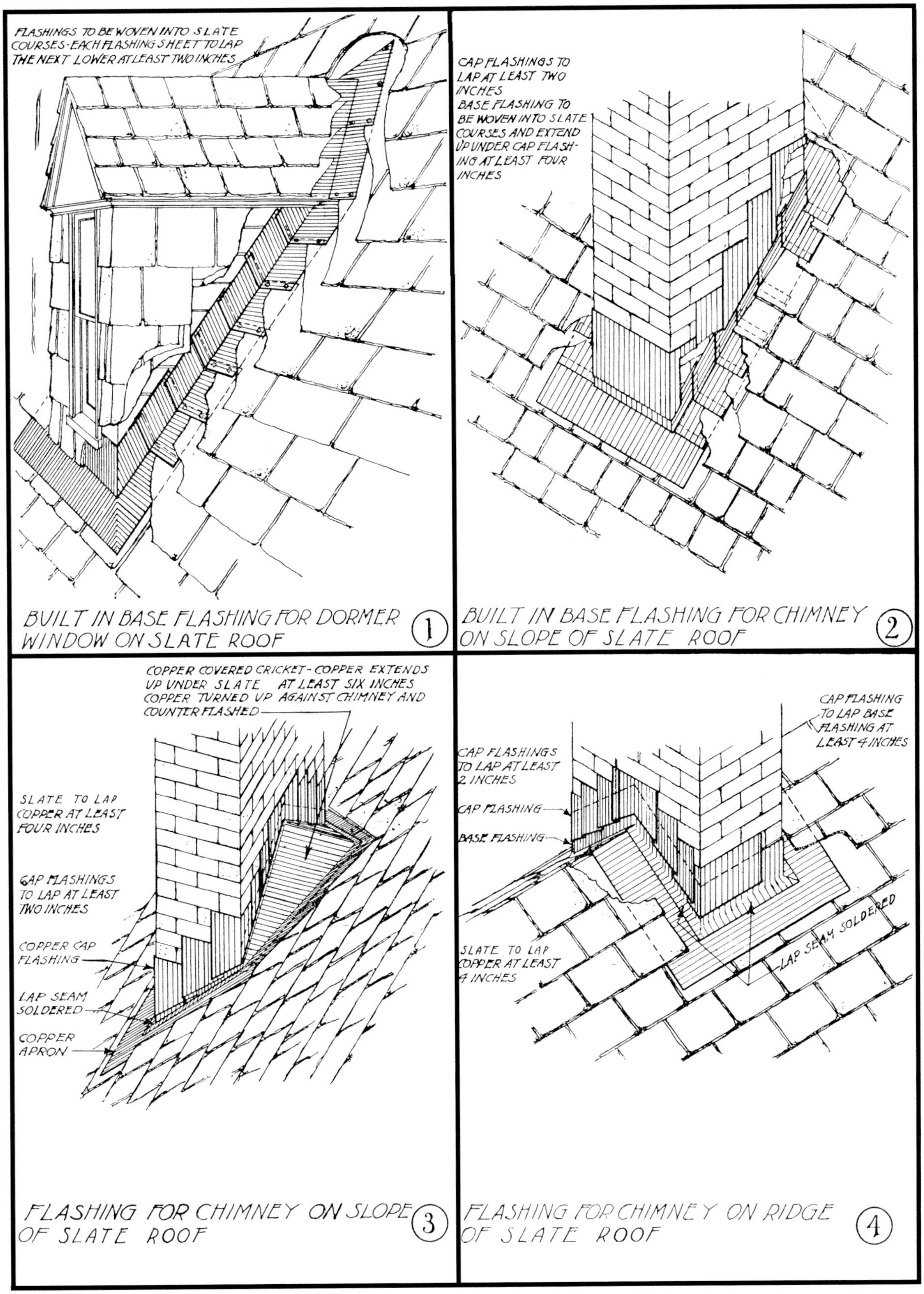

FIGURE 34. STANDARD DETAILS FLASHINGS

zinc is desired. The Association will be glad to furnish specifications for the painting of zinc upon request.

Rolled zinc for flashings should be not less than No. 11 zinc gauge (0.024″ thick) and should be laid in the usual manner, not nailed, but held in place by means of zinc clips or cleats. Zinc flashing against masonry, concrete and stucco should be laid on a good grade of water-proof sheathing paper. If the cap flashing is set in a reglet, it should be pointed with elastic cement.

Solder

The agent used to join pieces of metal into one length or sheet is known as "solder." The best grade, composed of equal parts of new tin and new lead, should be used and should conform to the "Standard Specifications for Solder Metal," Serial Designation B-32-21, American Society for Testing Materials. To hasten its melting and conserve the heat in the solder iron, a flux is used.

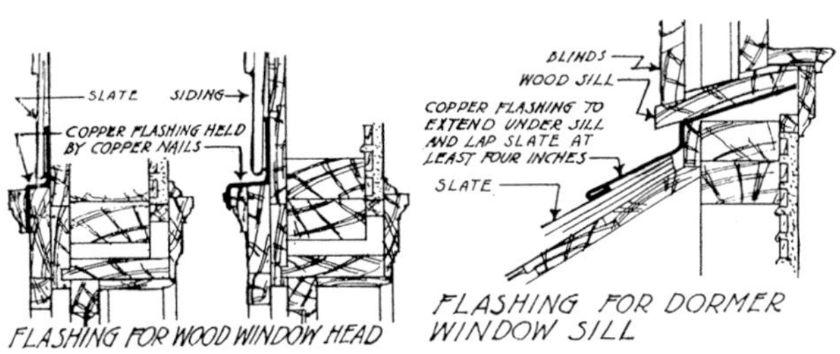

Figure 35

On copper, rosin is the best agent for this purpose, and the use of acid as a substitute should be avoided whenever possible. Acid flux of an improper kind may do irreparable damage to the finest workmanship. Rosin is harmless to the metal and makes good seams. There are some objections to its use, such as sloping roofs and windy days. Under these conditions it is much easier to use killed acid for it will not blow away although it may run down a slope and spatter on windy days. Rosin can be kept in place by "burning" it on with a small soldering copper just hot enough to melt the rosin. The proper preparation of acid for use as a flux is of greatest importance. The acid is hydrochloric or muriatic. Pieces of zinc are put in the quantity to be used until it stops working; then it is properly killed. If the killing is done hastily or by any one not familiar with the procedure, the acid may be used in a still active state and attack the copper. The acid to be used for the entire job should be prepared several days before the work starts and allowed to stand. On zinc and galvanized metal, acid should be used. Where the joints of the metal are not thoroughly sweated or soaked with solder, they may be loosened by expansion or contraction of the metal or leave small holes in the joint through which moisture readily finds its way.

A new product known as soldering salts may also be used as a flux. It is claimed that these salts do not require so hot a soldering iron and that they also have other advantages.

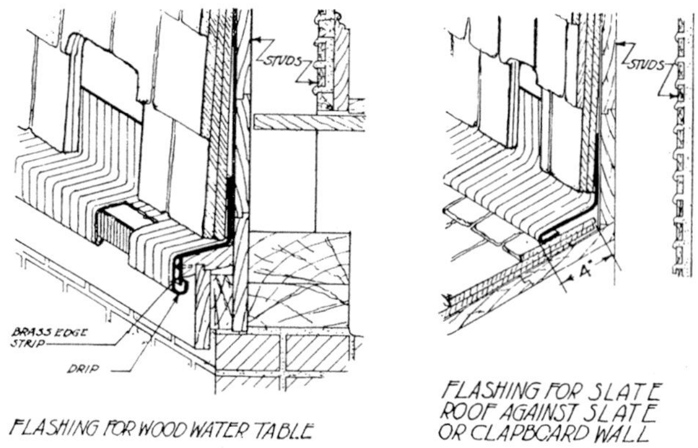

Figure 36

Galvanic Action

Dissimilar metals, when in contact in the presence of an electrolyte, set up galvanic action which results in the deterioration of the most electropositive metal.

Any possibility of galvanic action between copper and iron or steel should be carefully avoided by proper insulation. This insulation is effected in various ways, three of which are: (1) covering the steel member with asbestos, as is frequently done in skylight construction; (2) placing strips of sheet lead between the two metals, as when new copper gutters are placed in old iron hangers; and (3) heavily tinning the iron, as is often done with iron or steel gutter and leader supports.

Flashings—General

Flashing should be used at all intersections of vertical or projecting surfaces through the roof or against which the roof abutts, such as walls, parapets, dormers, sides of chimneys, etc. Flashings used over or under the roof covering and turned up on the vertical surface are known as "base flashings." Metal built into the vertical surface and bent down over the base flashing is termed a "cap flashing" or "counter flashing."

Base Flashings

The base flashings should be extended under the uppermost row the full depth of the slate or at least 4″ over the slate immediately below the metal. The vertical leg must be turned up not less than 4″ and preferably 8″ on the abutting surface. Where a vertical surface butts against the roof slope, it is necessary to build in the base

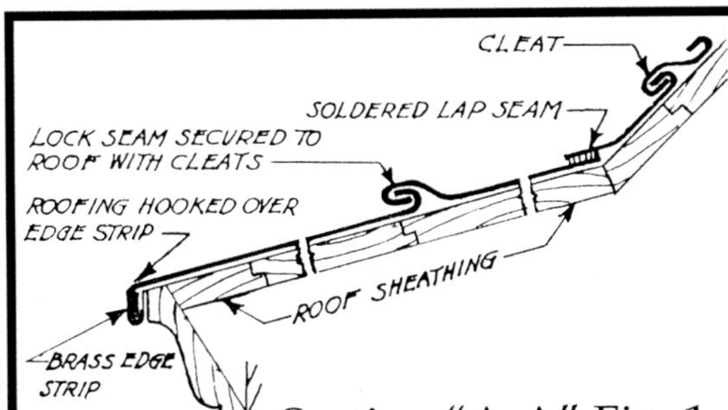

Section "A-A" Fig. 1

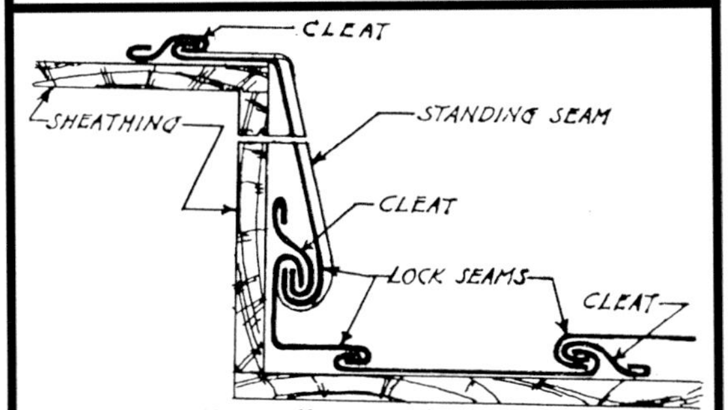

Section "B-B" Fig. 1

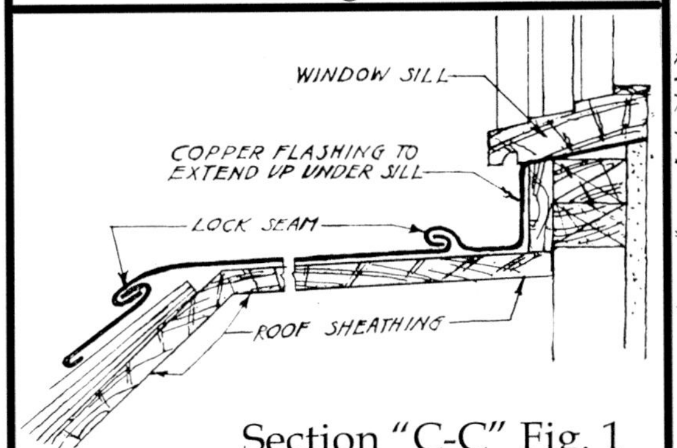

Section "C-C" Fig. 1

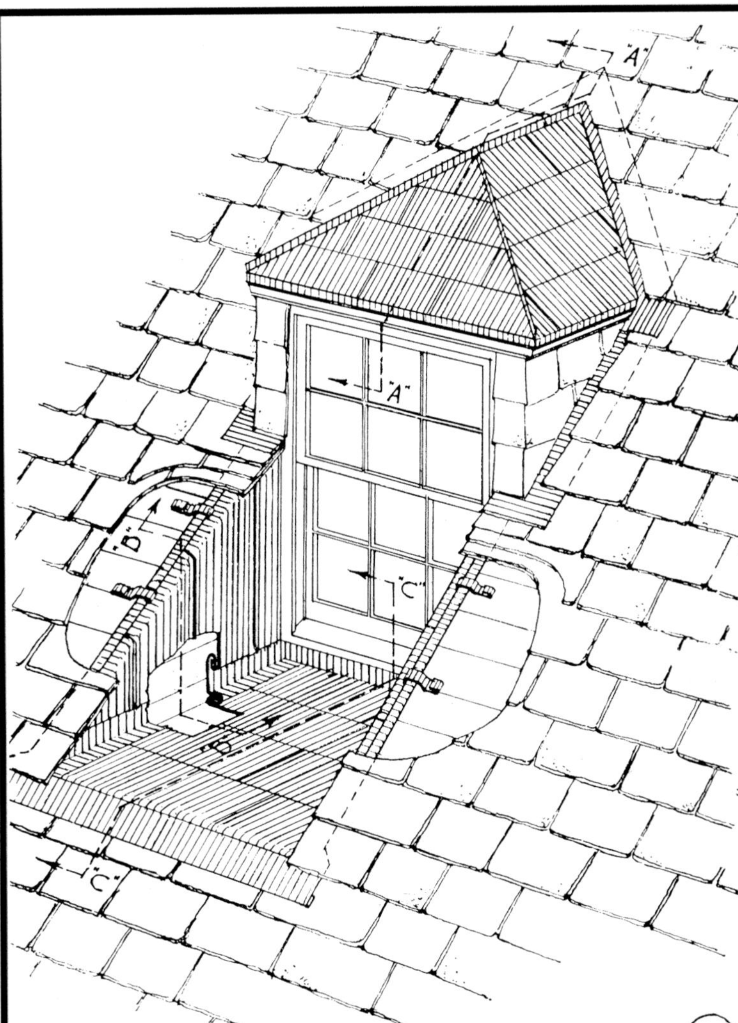

Flashing for a Recessed Dormer Window (1)

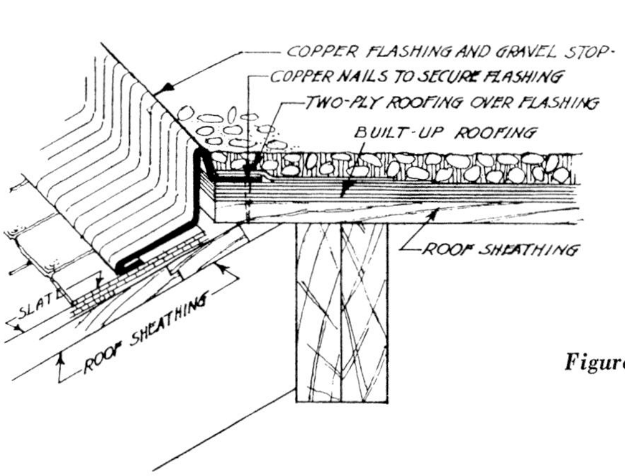

Figure 37

Flashing for Edge of Composition Deck Roof Above a Sloping Slate Roof

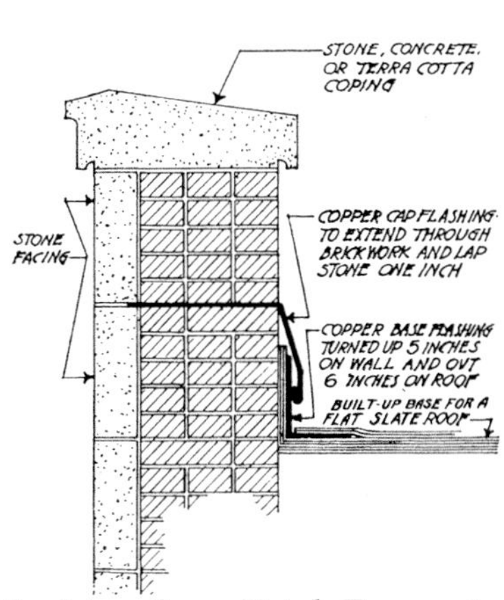

When a parapet wall is flashed on the top and back with metal, the flashing should be carried over and down to within one inch of bottom of cap flashing formed as here shown, so that most of the water is deflected out onto the slate.

Flashing for a Brick Parapet Wall Faced with Stone

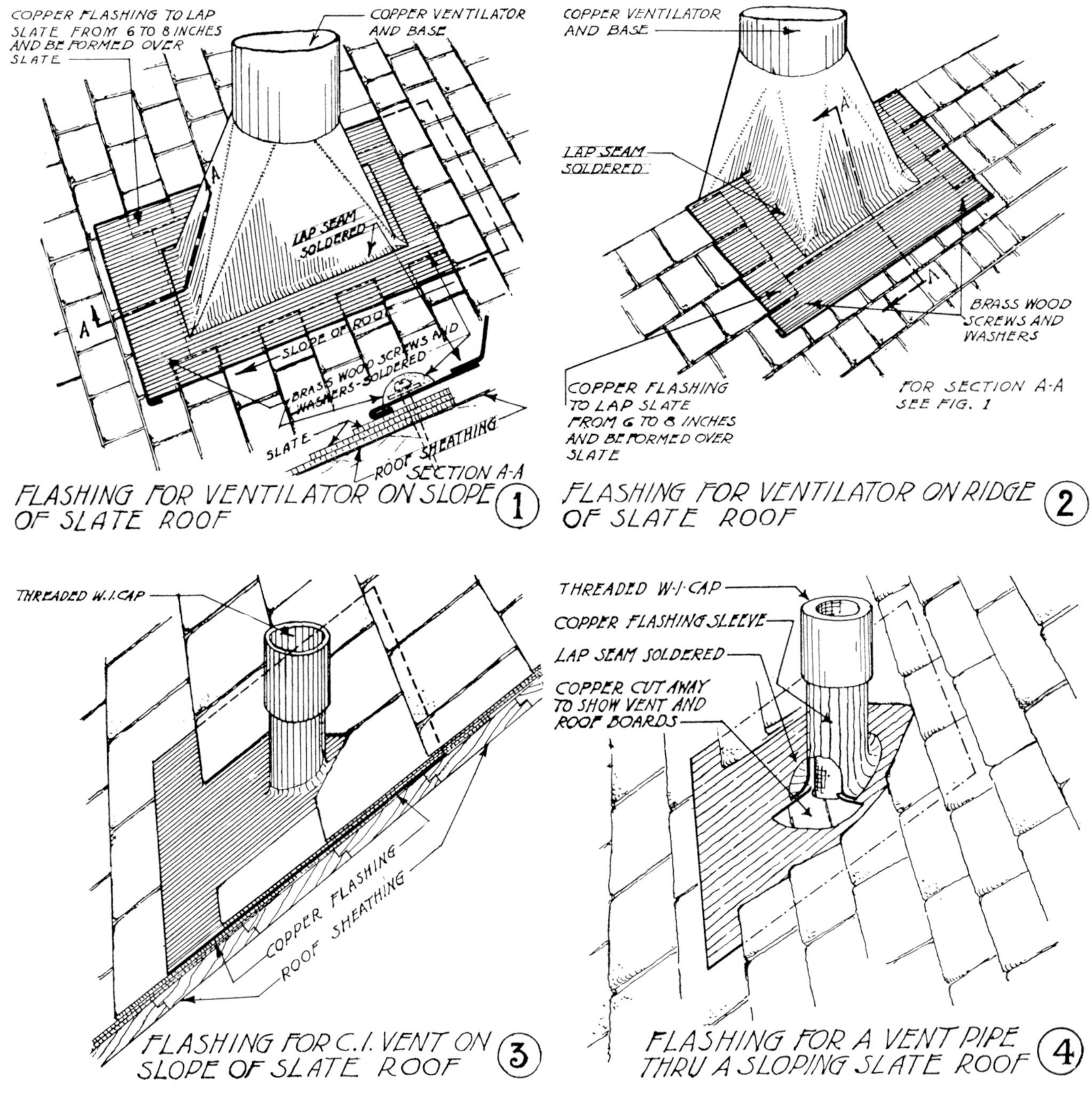

Figure 38

flashing with each course of slate as laid. Turn out 4″ on the slate and at least 8″ above the roof. If the roof stops against a stuccoed wall, a wood strip 4″ wide having a beveled top edge should be secured to the wall. The base flashing is then turned out over the slate at least 4″ and bent up vertically at least 3″ on the board. Except in unusual cases, it will be found satisfactory to turn the base flashing out 4″ on the roof surface and up on the vertical surface from 6″ to 8″ for either sloping or flat roofs. Posts, flagpoles, scuttles, etc., where projecting through the roof, should have base flashings. Vent pipes should have base flashings in the form of special sleeves or one of the numerous patented roof flashing devices.

Cap Flashings

Where the base flashing is not covered by vertical slate, siding, etc., a cap flashing must be used. This member should be built into the masonry joints not less than 2″, extend down over the base flashing 4″, and the edge bent back and up ½″. Reglets in stone or concrete are usually about 1″ wide and 1″ deep. The flashing should be formed and laid in the bottom of the cut and thoroughly caulked with molten lead on flat surfaces or lead wool on upright work. After caulking, the reglet is filled to the surface with elastic cement. Flashing hooks should be used to secure stepped flashings and the vertical legs be made tight with roofers' cement colored the same as the masonry. On the best work these flashings should be soldered.

It will be noted on the drawings that all exposed and unfastened flashings have the edge of the strip turned under ½″. This is done to give the strip stiffness against wind. Thus the sheet is held in place and the packing in of snow under the flashing is prevented.

Saddles or Crickets

Where a chimney or other vertical surface breaks through the roof at a right angle to the slope, a saddle or cricket must be built to throw the water away from the back of the vertical member. If the roof construction is of wood, use light rafter construction covered with sheathing boards, paper and sheet metal. If of very large area and exposed to prominent view, it should be slated the same as the other roof areas. Valleys will be formed with the main roof and it is recommended that they be of the open type. The size of the saddle is largely determined by the roof condition. It is usually sufficient to make the slope of the saddle the same as the roof.

It is most important that the saddle or cricket be of adequate size, of ample slope and well flashed as shown in Figure 34.

Estimating Flashings

In case it is desired to estimate the amount of metal required for flashing mitred hips (where metal is used), closed valleys, cheeks or side walls and base and cap flashings for walls, the following rules used by one roofing contractor may prove of value. This rule considers that flashings for hips and valleys are bent diagonally from corner to corner, while those for cheek or side walls are bent lengthways, allowing 4″ to turn up on the side wall and 4″ flashed under the slates.

RULE FOR ESTIMATING FLASHING

for Mitred Hips, Closed Valleys, Cheeks or Side Walls and Walls.

Hip and Valley Slip Flashings

Multiply the number of lineal feet of mitred hips and closed valleys by the following percentages, and the result will be the number of square feet of sheet metal for flashings, using slips.

Size of Slates	Size of Slips	Multiply by
12″	9 x 9	.800
14″	10 x 10	.825
16″	11 x 11	.858
18″	12 x 12	.908
20″	13 x 13	.946

Cheek or Side Wall Flashings

Size of Slates	Size of Slips	Multiplied by
12″	8 x 7½	1.111
14″	8 x 8½	1.031
16″	8 x 9½	.975
18″	8 x 10½	.934
20″	8 x 11½	.902

Wall Flashings

Multiply the number of lineal feet by 1.1, and it will give the number of square feet if flashings are 12″ wide.

If courses in valleys do not line up, it will require double the number of slips to flash properly.

ROOF CONSTRUCTION

IT has sometimes been stated that heavier roof construction is required for slate than for roofs of somewhat lighter materials. This is not true—that is, when slate of the commercial standard thickness is used. Such slate comprises the vast majority of all slate used for roofing. Obviously, for a graduated slate roof, where very thick slate is specified, the roof construction should be designed to provide for the increased weight of the slate. But for the regular or standard slate roof, any roof construction which conforms to good engineering practice, and is suitable for other roofing material, is adequate. This fact should be fully recognized in the interest of truth, economy and conservation. The weight of the roof covering is an insignificant quantity when compared with the combined weights of rafters, sheathing, snow and wind pressure which must be considered for all roofs, and water saturation of certain roofings other than slate.

The allowances for the combined snow and wind load on roofs is usually established by local building codes or custom and will naturally be found to vary in different sections of the country. Certain localities are subject to higher winds or heavier snowfalls, or both, than others. This anticipated load is several times the combined weights of the materials which make up the roof construction and covering.

Building codes usually specify requirements for the live load; that is, any load other than the weight of the construction itself, such as snow, wind, etc., which must be figured for all flat and sloping roofs. The average for flat roofs (less than 20° slope), taken from the codes of fifteen representative cities, is 40 pounds per square foot. For roofs of more than 20° slope, this requirement averages 30 pounds per square foot.

The dead load of the roof construction consists of the rafters, sheathing, or roof lath, the roof covering, and plastering on under side of rafters when it occurs. Rafters, depending upon size, spacing and kind of wood, may vary from 1 to 5 pounds per square foot of roof area. Sheathing boards will weigh about 2½ pounds and roof lath from 1¼ to 1½ pounds per square foot.

The roof covering may vary from 1 pound per square foot for three-ply ready roofing to 14 pounds for clay tile shingles. Wooden shingles, dry on the roof, average 2½ pounds per square foot. Slate of commercial standard thickness, dry or wet, will average 7 pounds per square foot when laid.

To demonstrate that standard thickness slate does not require heavier roof construction than is necessary for other roof coverings, the following comparison has been made, based on a typical roof:

	Wood Shingle Lbs.	Standard Slate Lbs.
Roof Covering......	2.5	7.0
Sheathing..........	2.5	2.5
Rafters............	3.0	3.0
Dead Load.........	8.0	12.5
Average Live Load required by law...	40.0	40.0
	48.0	52.5

Engineering calculations for the same roof conditions show that on this slight difference in weight, rafters of commercial stock size dictated by good practice for lighter materials will be of strength amply sufficient to accommodate commercial standard (3/16″) slate, whether on a new roof or for re-roofing purposes.

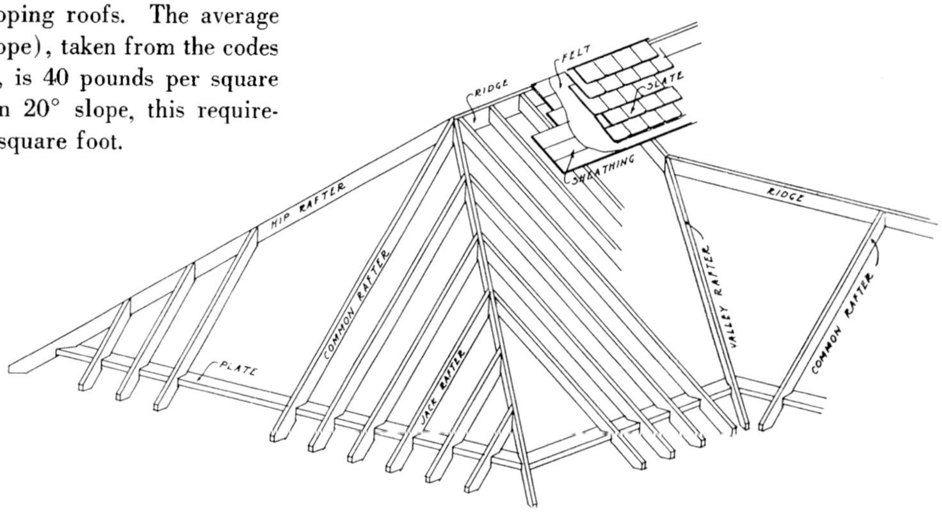

Figure 39. Parts of Roof Construction

HANDY RAFTER TABLES

The following tables provide a handy means of determining the size of roof rafters for any slope, uniformly loaded, for given spans in feet, and rafter spacings measured in inches from center to center.

Tables of sizes are based on the dead load, which includes the weight of the rafter, weight of sheathing (2.5 pounds per square foot) and weight of roof covering (8 pounds per square foot) and live loads of 50 and 30 pounds per square foot of roof surface considered as acting normal to the surface.

Where the ceilings are covered with some hard, inelastic material such as plaster, the span lengths should be limited by the deflection to prevent cracks. Where the ceilings are not so covered and where a small amount of sag or spring is not objectionable, the span length may be determined by the bending strength of the member instead of by its stiffness.

Allowable stresses for timber are usually prescribed in local building codes, but when they are not it is recommended that the values given in Table No. 4 be used. These values are taken from the recommendations of the Forest Products Laboratory, Department of Agriculture, at Madison, Wisconsin, that were officially adopted by the American Society for Testing Materials and the American Railway Engineering Association.

The following tables are compiled from "Wood Construction Information" issued September 1, 1922, by the National Lumber Manufacturers' Association.

TABLE No. 4.—Allowable Unit Stresses for Structural Timber
(Pounds per square inch)
Do not use this table if local Building Code specifies timber stresses.

Species of Timber	Modulus of Elasticity (E) (For determination of Deflection)	Stress in Extreme Fiber (f) (For determination of Bending)
Cedar, Western Red	1,000,000	900
Chestnut	1,000,000	950
Cypress	1,400,000	1,300
Douglas Fir (No. 1 Struct.)	1,600,000	1,600
Douglas Fir (No. 2 Struct.)	1,500,000	1,300
Douglas Fir, Rocky Mountain Region	1,200,000	1,100
Fir, Balsam	1,000,000	900
Gum, Red	1,200,000	1,100
Hemlock, Western	1,400,000	1,300
Hemlock, Eastern	1,100,000	1,000
Larch, Western	1,300,000	1,200
Maple, Sugar or Hard	1,600,000	1,500
Maple, Silver or Soft	1,100,000	1,000
Oak, White or Red	1,500,000	1,400
Pine, Southern Yellow (Dense)	1,600,000	1,600
Pine, Southern Yellow (Sound)	1,500,000	1,300
Pine, Eastern White	1,000,000	900
Pine, Western White	1,000,000	900
Pine, Norway	1,200,000	1,100
Redwood	1,300,000	1,200
Spruce, Red, White or Sitka	1,200,000	1,100
Tamarack, Eastern	1,300,000	1,200

TABLE No. 5.
Order of Strength of Standard Sizes of Rafters as Determined by their Modulus of Elasticity (E) and their Extreme Fiber Stress (f) in Pounds per square inch.

Deflection (E)	Bending (f)
2 x 4	2 x 4
2 x 6	2 x 6
3 x 6	3 x 6
2 x 8	2 x 8
3 x 8	2 x 10
2 x 10	3 x 8
3 x 10	2 x 12
2 x 12	3 x 10
3 x 12	2 x 14
2 x 14	3 x 12
3 x 14	3 x 14

TABLE No. 6.—Rafter Sizes for Roof of any Slope—Uniformly Loaded

Based on Modulus of Elasticity (E) and Limited by a Deflection of 1/360 of the Span.
Live Load—50 pounds per square foot.

(Use the following table when the roof is desired to be very rigid in order to prevent cracks in plaster or other inelastic material on the ceiling.) Use Modulus of Elasticity (E) given in building code—if not given in code use value for E for kind of wood used in table No. 4.

Span Between Plate and Ridge or Supporting Members in Feet	Rafter Spacing Center to Center in Inches	Size of Rafters (Nominal) in Inches			
		E=1,000,000	E=1,200,000	E=1,400,000	E=1,600,000
6	12	2 x 6	2 x 6	2 x 6	2 x 6
	16	2 x 6	2 x 6	2 x 6	2 x 6
	24	2 x 6	2 x 6	2 x 6	2 x 6
8	12	2 x 6	2 x 6	2 x 6	2 x 6
	16	3 x 6	2 x 6	2 x 6	2 x 6
	24	2 x 8	3 x 6	3 x 6	2 x 8
10	12	2 x 8	3 x 6	3 x 6	3 x 6
	16	2 x 8	2 x 8	2 x 8	3 x 6
	24	3 x 8	2 x 10	3 x 8	2 x 8
12	12	3 x 8	3 x 8	2 x 8	2 x 8
	16	2 x 10	3 x 8	3 x 8	3 x 8
	24	3 x 10	3 x 10	2 x 10	3 x 8
14	12	2 x 10	2 x 10	3 x 8	3 x 8
	16	3 x 10	2 x 12	2 x 10	2 x 10
	24	3 x 12	3 x 10	3 x 10	3 x 10
16	12	3 x 10	3 x 10	3 x 10	2 x 10
	16	3 x 12	3 x 10	3 x 10	3 x 10
	24	3 x 12	2 x 14	3 x 12	3 x 12
18	12	3 x 12	3 x 12	3 x 10	3 x 10
	16	3 x 12	3 x 12	3 x 12	2 x 12
	24	3 x 14	3 x 14	2 x 14	3 x 12
20	12	2 x 14	3 x 12	3 x 12	3 x 12
	16	3 x 14	3 x 14	3 x 12	3 x 12
	24	....	....	3 x 14	3 x 14
22	12	3 x 14	3 x 14	2 x 14	2 x 14
	16	....	3 x 14	3 x 14	3 x 14
	24	....	....	....	....
24	12	....	3 x 14	3 x 14	3 x 14
	16	....	....	....	3 x 14
	24	....	....	....	....

NOTE:—LIVE LOAD = 50 lbs. per square foot considered as acting normal to surface.
DEAD LOAD = weight of roof joist.
weight of sheathing (2.5 pounds per square foot).
weight of covering (8 pounds per square foot).

TABLE No. 7.—Rafter Sizes for Roof of any Slope—Uniformly Loaded.

Based on Extreme Fiber Stress (f)

Live Load—50 pounds per square foot.

(Use the following table when a small amount of sag or spring does not matter. Use Extreme Fiber Stress (f) given in local building code—if not given in code, use value for (f) for kind of wood used in Table 4.)

Span Between Plate and Ridge or Supporting Members in Feet	Rafter Spacing Center to Center in Inches	Size of Rafters (Nominal) in Inches									
		f=900	f=1000	f=1100	f=1200	f=1300	f=1400	f=1500	f=1600	f=1700	f=1800
6	12	2 x 6	2 x 6	2 x 6	2 x 6	2 x 6	2 x 6	2 x 6	2 x 6	2 x 6	2 x 6
	16	2 x 6	2 x 6	2 x 6	2 x 6	2 x 6	2 x 6	2 x 6	2 x 6	2 x 6	2 x 6
	24	2 x 6	2 x 6	2 x 6	2 x 6	2 x 6	2 x 6	2 x 6	2 x 6	2 x 6	2 x 6
8	12	2 x 6	2 x 6	2 x 6	2 x 6	2 x 6	2 x 6	2 x 6	2 x 6	2 x 6	2 x 6
	16	3 x 6	2 x 6	2 x 6	2 x 6	2 x 6	2 x 6	2 x 6	2 x 6	2 x 6	2 x 6
	24	3 x 6	3 x 6	3 x 6	3 x 6	3 x 6	2 x 6	2 x 6	2 x 6	2 x 6	2 x 6
10	12	3 x 6	3 x 6	2 x 6	2 x 6	2 x 6	2 x 6	2 x 6	2 x 6	2 x 6	2 x 6
	16	2 x 8	3 x 6	3 x 6	3 x 6	3 x 6	3 x 6	2 x 6	2 x 6	2 x 6	2 x 6
	24	2 x 10	2 x 10	2 x 10	2 x 10	2 x 8	3 x 6	3 x 6	3 x 6	3 x 6	3 x 6
12	12	2 x 8	3 x 6	3 x 6	3 x 6	3 x 6	3 x 6	3 x 6	2 x 6	2 x 6	2 x 6
	16	2 x 10	2 x 10	2 x 10	2 x 8	3 x 6	3 x 6	3 x 6	3 x 6	3 x 6	2 x 8
	24	2 x 12	2 x 12	2 x 10	2 x 10	2 x 10	2 x 10	2 x 10	2 x 10	2 x 10	2 x 8
14	12	2 x 10	2 x 10	2 x 10	2 x 10	2 x 8	3 x 6	3 x 6	3 x 6	3 x 6	3 x 6
	16	2 x 12	3 x 8	2 x 10	2 x 10	2 x 10	2 x 10	2 x 10	2 x 8	2 x 8	3 x 6
	24	3 x 10	3 x 10	2 x 12	2 x 12	2 x 12	2 x 12	2 x 10	2 x 10	2 x 10	2 x 10
16	12	2 x 12	2 x 10	2 x 10	2 x 10	2 x 10	2 x 10	2 x 10	2 x 8	2 x 8	3 x 6
	16	3 x 10	2 x 12	2 x 12	2 x 12	3 x 8	3 x 8	2 x 10	2 x 10	2 x 10	2 x 10
	24	3 x 12	2 x 14	2 x 14	3 x 10	3 x 10	2 x 12	2 x 12	2 x 12	2 x 12	2 x 12
18	12	2 x 12	2 x 12	2 x 12	2 x 12	2 x 10	2 x 10	2 x 10	2 x 10	2 x 10	2 x 10
	16	2 x 14	2 x 14	3 x 10	2 x 12	2 x 12	2 x 12	2 x 12	3 x 8	2 x 10	2 x 10
	24	3 x 14	3 x 14	3 x 12	2 x 14	2 x 14	2 x 14	3 x 10	3 x 10	2 x 12	2 x 12
20	12	2 x 14	3 x 10	2 x 12	2 x 12	2 x 12	2 x 12	2 x 12	2 x 10	2 x 10	2 x 10
	16	3 x 12	2 x 14	2 x 14	2 x 14	3 x 10	3 x 10	2 x 12	2 x 12	2 x 12	2 x 12
	24	....	3 x 14	3 x 14	3 x 14	3 x 12	3 x 12	2 x 14	2 x 14	2 x 14	2 x 14
22	12	2 x 14	2 x 14	2 x 14	3 x 10	3 x 10	2 x 12	2 x 12	2 x 12	2 x 12	2 x 12
	16	3 x 14	3 x 14	3 x 12	2 x 14	2 x 14	2 x 14	2 x 14	3 x 10	3 x 10	2 x 12
	24	....	....	....	3 x 14	3 x 14	3 x 14	3 x 14	3 x 12	3 x 12	2 x 14
24	12	3 x 14	3 x 12	2 x 14	2 x 14	2 x 14	3 x 10	3 x 10	2 x 12	2 x 12	2 x 12
	16	....	3 x 14	3 x 14	3 x 14	3 x 12	2 x 14	2 x 14	2 x 14	2 x 14	2 x 14
	24	....	....	....	....	....	3 x 14	3 x 14	3 x 14	3 x 14	3 x 14

NOTE:—LIVE LOAD =50 lbs. per square foot considered as acting normal to surface.
DEAD LOAD=weight of roof joist.
weight of sheathing (2.5 pounds per square foot).
weight of covering (8 pounds per square foot).

The Building Code Committee of the Department of Commerce, in "Minimum Live Loads Allowable for Use in Design of Buildings," November 1, 1924, states regarding roof loads:

"Roofs having a rise of 4 inches or less per foot of horizontal projection shall be proportioned for a vertical live load of 30 pounds per square foot of horizontal projection applied to any or all slopes. With a rise of more than 4 inches and not more than 12 inches per foot, a vertical live load of 20 pounds on the horizontal projection shall be assumed. If the rise exceeds 12 inches per foot, no vertical live load need be assumed, but provision shall be made for a wind force acting normal to the roof surface (on one slope at a time) of 20 pounds per square foot of such surface. (See Appendix, par. 7.)"

This Appendix paragraph reads:

"The minimum roof loads specified in Part II, section 5, apply only in localities where snow loads are not an important consideration. Roofs having a slope of less than 4 inches per foot are always liable to accidental loading, such as groups of moving people, storage of material, etc. Hence the necessity of moderate unit loads even where snow is not to be expected. Where large snow loads are to be anticipated, the loadings prescribed should be increased in accordance with local experience."

In view of these recommendations, the National Slate Association has deemed it advisable to include tables for a live load of 30 pounds per square foot, and these follow.

TABLE No. 8.—Rafter Sizes for Roof of Any Slope—Uniformly Loaded.
Based on Modulus of Elasticity (E) and Limited by a Deflection of 1/360 of the Span.
Live Load—30 pounds per square foot.
(Use the following table when the roof is desired to be very rigid in order to prevent cracks in plaster or other inelastic material on the ceiling.) Use Modulus of Elasticity (E) given in building code—if not given in code use value for E for kind of wood used in table No. 4.

Span Between Plate and Ridge or Supporting Members in Feet	Rafter Spacing Center to Center in Inches	Size of Rafters (Nominal) in Inches.			
		E=1,000,000	E=1,200,000	E=1,400,000	E=1,600,000
6	12 16 24	2 x 4 2 x 6 2 x 6	2 x 4 2 x 6 2 x 6	2 x 4 2 x 4 2 x 6	2 x 4 2 x 4 2 x 6
8	12 16 24	2 x 6 2 x 6 3 x 6	2 x 6 2 x 6 2 x 6	2 x 6 2 x 6 2 x 6	2 x 6 2 x 6 2 x 6
10	12 16 24	3 x 6 2 x 8 2 x 8	2 x 6 3 x 6 2 x 8	2 x 6 3 x 6 2 x 8	2 x 6 2 x 6 3 x 6
12	12 16 24	2 x 8 3 x 8 2 x 10	2 x 8 2 x 8 3 x 8	3 x 6 2 x 8 3 x 8	3 x 6 2 x 8 3 x 8
14	12 16 24	3 x 8 2 x 10 3 x 10	3 x 8 3 x 8 3 x 10	2 x 8 3 x 8 2 x 10	2 x 8 3 x 8 2 x 10
16	12 16 24	3 x 10 3 x 10 3 x 12	2 x 10 3 x 10 2 x 12	3 x 8 2 x 10 3 x 10	3 x 8 2 x 10 3 x 10
18	12 16 24	3 x 10 3 x 12 2 x 14	3 x 10 2 x 12 3 x 12	3 x 10 3 x 10 3 x 12	2 x 10 3 x 10 3 x 12
20	12 16 24	3 x 12 3 x 12 3 x 14	2 x 12 3 x 12 3 x 14	3 x 10 3 x 12 3 x 12	3 x 10 2 x 12 3 x 12
22	12 16 24	2 x 14 3 x 14	3 x 12 3 x 14 3 x 14	3 x 12 3 x 12 3 x 14	2 x 12 3 x 12 3 x 14
24	12 16 24	3 x 14	3 x 14 3 x 14	3 x 12 3 x 14	3 x 12 3 x 14 3 x 14

NOTE:—LIVE LOAD =30 lbs. per square foot considered as acting normal to surface.
DEAD LOAD=weight of roof joist.
weight of sheathing (2.5 pounds per square foot).
weight of covering (8.0 pounds per square foot).

TABLE No. 9.—Rafter Sizes for Roof of Any Slope—Uniformly Loaded.

Based on Extreme Fiber Stress (f)

Live Load—30 pounds per square foot.

(Use the following table when a small amount of sag or spring does not matter. Use Extreme Fiber Stress (f) given in local building code—if not given in code, use value for (f) for kind of wood used in Table No. 4.)

Span Between Plate and Ridge or Supporting Members in Feet	Rafter Spacing Center to Center in Inches	Size of Rafters (Nominal) in Inches.									
		f=900	f=1000	f=1100	f=1200	f=1300	f=1400	f=1500	f=1600	f=1700	f=1800
6	12	2 x 4	2 x 4	2 x 4	2 x 4	2 x 4	2 x 4	2 x 4	2 x 4	2 x 4	2 x 4
	16	2 x 4	2 x 4	2 x 4	2 x 4	2 x 4	2 x 4	2 x 4	2 x 4	2 x 4	2 x 4
	24	2 x 6	2 x 6	2 x 6	2 x 6	2 x 4	2 x 4	2 x 4	2 x 4	2 x 4	2 x 4
8	12	2 x 6	2 x 6	2 x 6	2 x 4	2 x 4	2 x 4	2 x 4	2 x 4	2 x 4	2 x 4
	16	2 x 6	2 x 6	2 x 6	2 x 6	2 x 6	2 x 6	2 x 6	2 x 4	2 x 4	2 x 4
	24	3 x 6	2 x 6	2 x 6	2 x 6	2 x 6	2 x 6	2 x 6	2 x 6	2 x 6	2 x 6
10	12	2 x 6	2 x 6	2 x 6	2 x 6	2 x 6	2 x 6	2 x 6	2 x 6	2 x 6	2 x 4
	16	3 x 6	2 x 6	2 x 6	2 x 6	2 x 6	2 x 6	2 x 6	2 x 6	2 x 6	2 x 6
	24	2 x 8	3 x 6	3 x 6	3 x 6	3 x 6	3 x 6	2 x 6	2 x 6	2 x 6	2 x 6
12	12	3 x 6	3 x 6	2 x 6	2 x 6	2 x 6	2 x 6	2 x 6	2 x 6	2 x 6	2 x 6
	16	2 x 8	3 x 6	3 x 6	3 x 6	3 x 6	3 x 6	2 x 6	2 x 6	2 x 6	2 x 6
	24	2 x 10	2 x 10	2 x 10	2 x 8	2 x 8	3 x 6	3 x 6	3 x 6	3 x 6	3 x 6
14	12	2 x 8	3 x 6	3 x 6	3 x 6	3 x 6	3 x 6	2 x 6	2 x 6	2 x 6	2 x 6
	16	2 x 10	2 x 10	2 x 10	2 x 8	3 x 6	3 x 6	3 x 6	3 x 6	3 x 6	3 x 6
	24	2 x 12	3 x 8	2 x 10	2 x 10	2 x 10	2 x 10	2 x 10	2 x 10	2 x 8	3 x 6
16	12	2 x 10	2 x 10	2 x 10	2 x 8	3 x 6	3 x 6	3 x 6	3 x 6	3 x 6	3 x 6
	16	3 x 8	2 x 10	2 x 10	2 x 10	2 x 10	2 x 10	2 x 8	2 x 8	3 x 6	3 x 6
	24	3 x 10	2 x 12	2 x 12	2 x 12	3 x 8	3 x 8	2 x 10	2 x 10	2 x 10	2 x 10
18	12	2 x 10	2 x 10	2 x 10	2 x 10	2 x 10	2 x 10	2 x 8	3 x 6	3 x 6	3 x 6
	16	2 x 12	2 x 12	2 x 12	2 x 10	2 x 10	2 x 10	2 x 10	2 x 10	2 x 10	2 x 10
	24	2 x 14	2 x 14	3 x 10	2 x 12	2 x 12	2 x 12	2 x 12	2 x 12	2 x 10	2 x 10
20	12	2 x 12	2 x 12	2 x 10	2 x 10	2 x 10	2 x 10	2 x 10	2 x 10	2 x 10	2 x 8
	16	3 x 10	2 x 12	2 x 12	2 x 12	2 x 12	3 x 8	2 x 10	2 x 10	2 x 10	2 x 10
	24	3 x 12	2 x 14	2 x 14	2 x 14	3 x 10	3 x 10	2 x 12	2 x 12	2 x 12	2 x 12
22	12	3 x 10	2 x 12	2 x 12	2 x 12	3 x 8	2 x 10	2 x 10	2 x 10	2 x 10	2 x 10
	16	2 x 14	2 x 14	3 x 10	2 x 12	2 x 12	2 x 12	2 x 12	2 x 12	3 x 8	2 x 10
	24	3 x 14	3 x 14	3 x 12	2 x 14	2 x 14	2 x 14	2 x 14	3 x 10	3 x 10	2 x 12
24	12	2 x 14	3 x 10	2 x 12	2 x 12	2 x 12	2 x 12	2 x 12	2 x 10	2 x 10	2 x 10
	16	3 x 12	2 x 14	2 x 14	2 x 14	3 x 10	3 x 10	2 x 12	2 x 12	2 x 12	2 x 12
	24	....	3 x 14	3 x 14	3 x 14	3 x 12	3 x 12	2 x 14	2 x 14	2 x 14	2 x 14

NOTE:—LIVE LOAD = 30 lbs. per square foot considered as acting normal to surface.
DEAD LOAD = weight of roof joist.
weight of sheathing (2.5 pounds per square foot).
weight of covering (8.0 pounds per square foot).

It may also be pointed out that, although the wind pressure will remain the same for any roof no matter what the covering, Kidder's Architects and Builders Pocketbook gives slate a snow load credit of from 3 to 12 pounds per square foot over shingles, depending upon the slope of the roof and the climate.

CONSTRUCTION DETAILS

IN any type of construction, it is important that the structural members be well tied together. This is especially important in roof construction at the joint of the wall plate and rafter or a purlin and rafter. Ceiling joists or other horizontal members spiked to the foot of the rafters ordinarily accomplish this. In addition, the foot cut of the rafter should be so proportioned and designed that it will resist as much as possible the tendency to "slip" or spread as well as provide sufficient area to prevent crushing of the rafter or plate fibers. Good practice dictates that a wood plate on a masonry wall be bedded in mortar and securely anchored or bolted to the wall. It is needless to say that the walls receiving the roof load should be of sufficient strength and so built that the reactions of the roof and various floor loads will be kept well within or near the center of gravity of the wall. The footings should be properly proportioned to keep the load on the foundation within the soil bearing value.

At the ridge it is advisable to use a ridge pole or board, rather than butting the peak ends of rafters of opposite slopes. The use of a board or pole affords better rafter bearing and will provide a straighter ridge. Collar beams spiked to the rafters near the peak assist in resisting the tendency to spread. If the simple principle of triangles used in truss design is followed in tying and bracing the structural members, a safe and economical roof framing layout should result.

Sheathing or Roof Boards

Wood sheathing is commonly of tongued and grooved boards dressed on either one or both sides. If dressed on one side only, this side should be laid down. The boards should be 1″ nominal thickness (not less than 25/32″) and from 6″ to 10″ wide. A width of 8″ is a convenient average size and preferable to the 10″ width. The tongue should be laid toward the ridge. Square-edge, rough boards, when used, should not be over 8″ wide. Shiplap boards are preferred for sheathing by some roofers, who claim they have less tendency to warp than tongued and grooved when nailed too tightly together.

It is important that a smooth, solid job of sheathing be obtained to allow proper laying of the slate. Joints between the ends of boards should occur over the rafters and the ends should be securely nailed at these points. The boards should also be nailed with 10-penny nails at both edges to every rafter. If 10″ boards are used, it is advisable to also nail the center of the board to its bearing. Tongued and grooved boards should not be drawn up too close, as there is danger of the roof surface buckling if rained upon before the slate is laid.

The tongued and grooved boards are generally to be recommended as offering a smooth surface and a tighter and warmer roof than other types. Square-edged boards are liable to warp and curl, and although this is seldom sufficient to break the slate, it may raise the courses and mar the appearance of the roof. Uneveness of the sheathing surface invariably results in a noticeable uneveness in the roof surface. For this reason, scrap lumber, concrete form lumber and pieces of uneven thickness should not be permitted as a covering for the rafters. Improperly nailed boards and loose joints make it difficult to lay the slate for the process of driving the slating nails into the sheathing springs and loosens the adjoining slate.

Roof lath or strips 1″ thick and 2″ or 3″ wide and without felt are used in many localities. The spacing of the lath will vary with the length of the slate used. The upper end of each slate should rest on the center of strip and the slate should be so punched that the nails will be driven into a lath. When the slate is laid with the standard 3″ lap, the lath should be spaced as follows:

Length of slate (Inches)	Spacing of lath (Inches)	Length of slate (Inches)	Spacing of lath (Inches)
24	10½	16	6½
22	9½	14	5½
20	8½	12	4½
18	7½		

This method of supporting the slate will not provide the warmth and the resultant saving in heating costs of other methods. While suitable for barns and similar structures, it is not recommended for use upon residences and important buildings except under favorable climatic conditions.

Round Valley Foundation

The foundation for the round valley is part of the sheathing work. There are three methods which may be used. The first, suitable for valleys of slight curvature, consists of a wide board (usually 12″) made with tapering sides and nailed into the angle formed by the intersecting roofs.

In the second method, suitable for any curvature or radius desired, 3″ blocks cut to fit the valley angle and sawed to the proper radius are nailed over the roof sheathing and spaced approximately the same as the exposure of the slates. The blocks thus form nailing strips under the slate. The size of the blocks will vary, due to the diminishing size of the valley as it approaches the ridge.

Unusual Slate and Glass Roof Baseball Cage, Amherst College

The third and usually the most satisfactory method is a combination of the first two. The 3″ nailing blocks cut to fit the valley angle and curvature of the valley are spaced from 20″ to 30″ apart over the regular sheathing. Tapered strips 7/8″ x 2″ or 3″ wide are nailed over the blocks and lengthwise of the valley.

The important consideration in any of these methods is a solid foundation of accurate form to support the slates and establish the desired shape of the valley.

Fireproof Construction

For fireproof construction, a nailing concrete or gypsum slab are sometimes used, into which the slating nails are driven. The materials selected for this purpose should be given careful consideration as to their nail-holding power and durability. When a roof slab is not used, the slates are wired or otherwise secured to angle iron structural members. The slates are punched with four holes, wire run through each set of two holes and around the angle. The ends of the wire are then twisted to fasten the slate tightly to the angle. (See Figure 40.)

Fireproof Construction to Receive Slate, Bennett Hall, University of Pennsylvania

Special fasteners are sometimes employed to attach the slate to the angle. As the angles must be spaced the same as the exposure of the slates, long slates are to be preferred as they reduce the number of angles required.

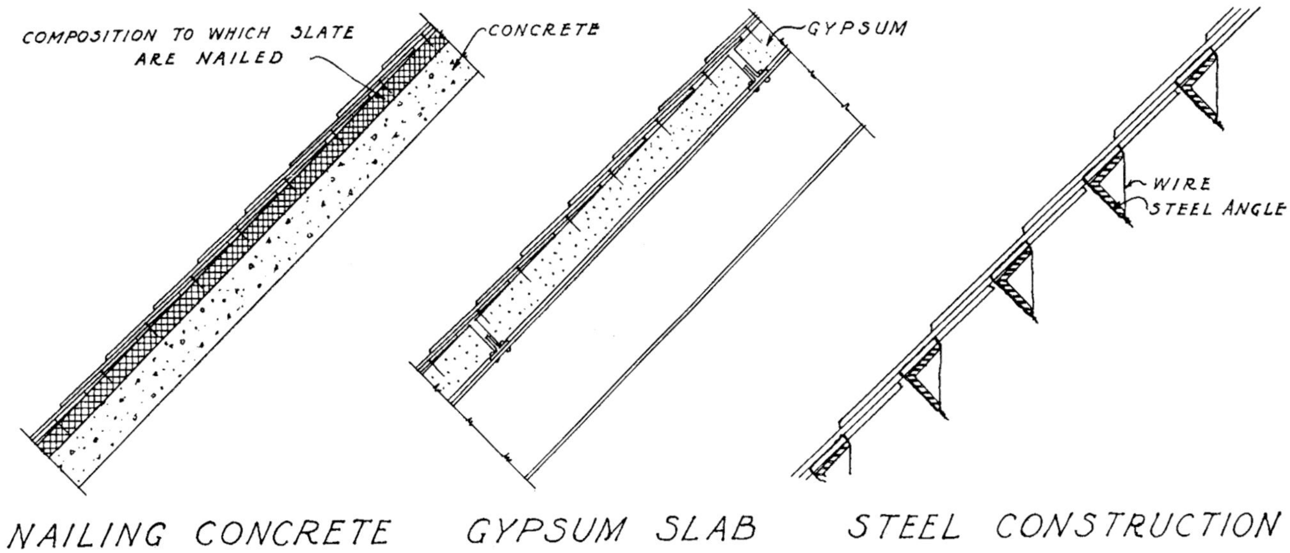

Figure 40. Slate on Various Types of Fireproof Construction

THE slope of a roof has been defined as the angle of inclination that the roof makes with a horizontal plane. The effect of this slope on the lap of the slate is shown in Figure 41.

There are three methods of describing the slope of any roof. First, in terms of vertical rise in inches to each foot of horizontal run, as "1 inch to the foot." Second, in terms of ratio of the total rise of the roof to its total span, as "¼ pitch or slope"; that is, the height of the roof is equal to ¼ of its total span. Third, in terms of degrees and minutes of the inclination of the roof to the horizontal plane.

The slope is dependent upon climatic conditions and the design, and determines the method of laying the slate and the lap required. For further information, see page 15.

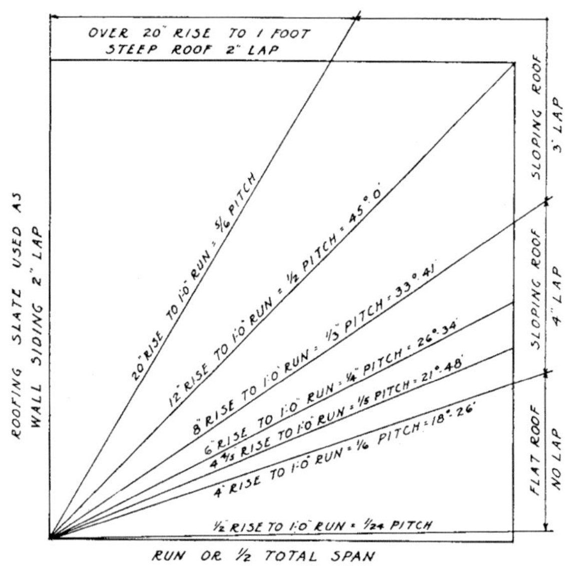

Figure 41. Lap of Slate for Various Roof Slopes

Eaves Troughs, Gutters and Conductor Pipes

Upon the proper proportioning of the gutters and conductors or leaders will often depend the satisfactory performance of the roof.

The size of leaders and gutters depends upon the amount of rainfall and the number of leaders used. One rule much in use is to provide 1 square inch of leader area for every 150 square feet of roof area. Leaders should not be spaced more, and preferably less, than 75′ apart for flat roofs and 50′ apart for sloping roofs.

All gutters should have a pitch of about 1″ in 16′ in order that they may be washed clean during rainfall. Gutters made 2″ larger than the leaders will permit a more practical connection than if made the same size.

When more than 50′ apart, the leaders should be increased 1″ for each additional 20′ of space between leaders for sloping roofs and for each 30′ for flat roofs.

For more detailed information, refer to Part 3 of "Copper Flashings" handbook of the Copper and Brass Research Association, 25 Broadway, New York City.

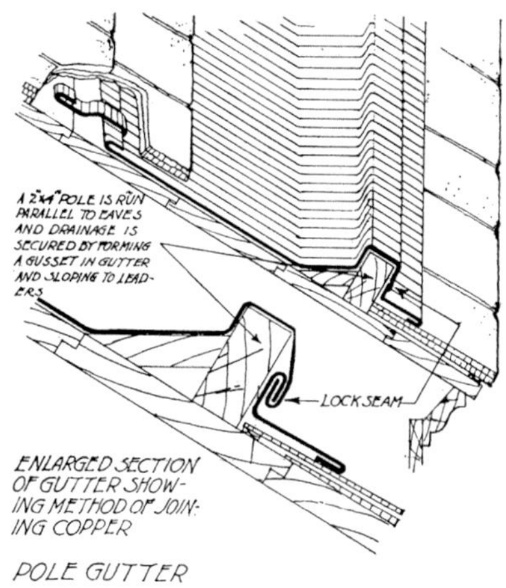

Figure 42

Numerous kinds of gutters are in use, among them being the "hanging gutter," which includes the "eaves trough" and the "moulded gutter"; the "box gutter," also known as the "built-in" or "lined gutter"; several types of "standing gutters," known by various names, and the "sunk gutter," which is also sometimes called the "box" or "built-in gutter." Each of these has its suitable use and location, as well as its disadvantages.

The "eaves trough" is generally made in semi-circular section and hung from the eaves by metal hangers which are adjustable to regulate the gutter slope. "Moulded gutters" are set true with the eaves line and are sometimes made with a sloped inner lining in order not to detract from the appearance of the building.

There are many types of hangers for "eaves troughs" or "hanging gutters." It is advisable to select a type of hanger not fastened directly under the slate, as the weight of snow or ice collecting in the gutter may pry up to the hanger and break the slate.

Hangers should be placed not further than 2′ 6″ apart, because of the liability of the gutter sagging when insufficiently supported. They should never be fastened rigidly to the gutter, as this will prevent the expansion and contraction of the metal.

The "box gutter" is most suitable for a building having a wooden cornice. It is in form a metal-lined box, the bottom of which is sloped. The metal lining should be carried over the eave mold and up the slope sufficiently far to prevent damage if the gutter or outlet become clogged and water is not carried off promptly.

One type of "standing gutter" is shown in Figure 42. In any type a cant strip should be placed under the lower courses of slate so as to form a drip at the edge and prevent water from being drawn up under the slate by capillary attraction and passing over the edge of the flashing.

"Sunk gutters" are built below the surface of the roof and are lined with metal. They are closed at the ends so the crown mold can be run up to the gable. This type

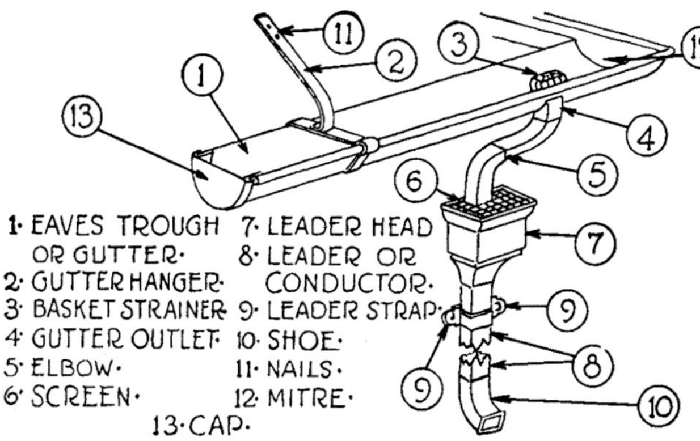

Figure 43. Parts of a Gutter

of gutter, as well as the "standing gutter," should always be placed as close as possible to the wall line of the house in order that the heat from the inside of the house may help melt any snow or ice which may form. The lining should be carried well up the slope.

When gutters are placed behind parapet walls, the gutter and flashing should be formed of a continuous piece of metal, which should extend, if practicable, to the top of the coping and up under the slate to a point at least 3″ above the top of the coping. If the gutter lining cannot be carried up to the top of the coping, it should be turned over and inserted in a reglet at a height of not less than 12″. It should be well wedged into the block with lead wool. The height to which the metal should be carried will be determined by the roof area drained, its slope, etc. On roofs of this character, it is essential that provision should be made for the escape of water if the leaders do not work. If the water is allowed to collect, it will not only cause a heavy and perhaps dangerous load, but may also work its way over the flashings and down into the building. Scuppers large enough to preclude any possibility of clogging (at least 4″ by 12″) should be provided and should be unobstructed by screens or other devices.

For inside leaders, a good grade of heavy cast iron or galvanized wrought iron pipe should be used and a trap provided when such leaders open near dormer windows or ventilating shafts. Outside leaders should be made of non-ferrous metal. In climates where freezing occurs, a rectangular-shaped leader or one having corrugations is to be preferred to a smooth circular section, as the former provides room for expansion if ice forms in the leader pipe.

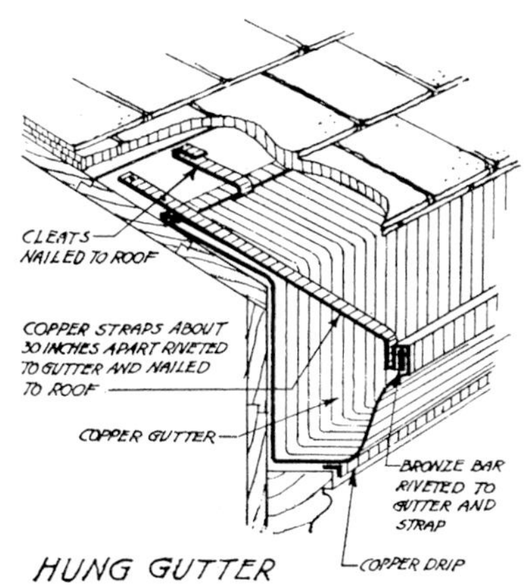

Figure 44

Leaders are often made with ornamental heads, and when used, the goose necks or offsets should empty into the head and not be joined to the downspout. If more than one fall of gutter empties into the leader, a head should always be provided. This should be of suitable dimensions to accommodate the flow of water into it.

With any type of gutter it is essential that the water be led away as quickly as possible. This is especially important in a climate where ice forms readily, as the snow will melt on the warm part of the roof, run on to the colder part and form ice in the gutter unless carried away at once. The ice will sometimes be heavy enough to bend or break the gutter and will almost always clog the downspout and back water over the edges of the gutter or on to the roof and under the roofing material.

Provision should always be made against the choking of leaders or outlets by falling leaves or other débris, and strainers of wire or heavy metal, depending upon the flow of water and roof area, installed. Gutters and leaders should be inspected at least twice a year to remove leaves, rubbish, and repair or repaint if necessary. The fastenings of the gutter should be closely examined and strainers renewed when required.

Leaders from higher or larger roofs should never be permitted to discharge water on to lower or smaller roofs, if such an arrangement can possibly be avoided. It is recommended that each gutter be individually drained down to the ground. Sometimes it is necessary to run the leader from a higher roof through the porch

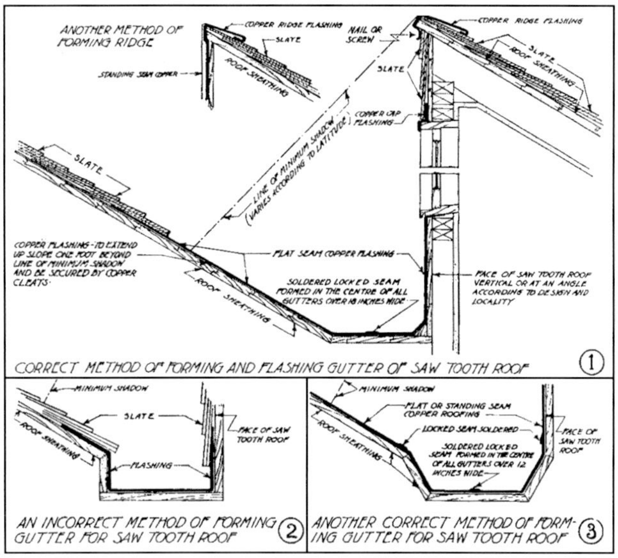

Figure 45

roof to accomplish this, or carry the leader across the porch roof to the gutter. In the latter case the leader should never have a slope of less than 45°; otherwise, ice may form in the leader. When it does become absolutely necessary to discharge the water on to a lower roof, metal spreaders should be used.

When a mansard roof is of slate and the deck is copper covered, it is sometimes advisable to frame a gutter in the deck because the drippings from the deck may discolor the slate. This gutter should be lined with cold rolled sheet copper and have a water outlet lead to the inside of the building.

When a low roof drains into a leader from a high roof, it is sometimes advisable to install a trap in the lower roof leader. If this is not done, the water pouring down from the higher roof may back up on to the low roof.

Adequate provision should always be made for carrying the water discharged by the leaders away from the walls of the building. This can sometimes be accomplished by "splash stones" or masonry gutters, or if the leaders do not connect with storm sewers, a cistern for rain water may be dug not closer than 50′ to the building. This may be lined to collect the water for household use or left unlined in order that the water may seep into the soil. The drain from leader to sewer or cistern should be placed below the frost line and slope about 1″ in 10′.

Simplified Sizes and Weights

In accordance with the unanimous action on October 14, 1924, of the general conference of representatives of manufacturers, distributors and users of eaves trough and conductor pipe, the United States Department of Commerce, through the Bureau of Standards, recommends that simplified rules, practices, sizes and weights of eaves trough and conductor pipe be established as follows:

Plain round conductor pipe: 2″, 3″, 4″, 5″ and 6″.
Round corrugated conductor pipe: 2″, 3″, 4″, 5″ and 6″.
Square corrugated conductor pipe: 2″, 3″, 4″ and 5″.
Eaves trough: 3½″, 4″, 5″, 6″, 7″ and 8″.
Conductor pipe elbows: No. 1, 45°; No. 2, 60°; No. 3, 75°; No. 4, 90°.

1. Along with the elimination of certain sizes of conductor pipe and eaves trough goes also that of the fittings formerly used therewith.

2. No eaves trough or conductor pipe to be made lighter than 28-gauge full weight; 27-gauge is to be eliminated.

3. All elbows, shoes, mitres and all accessories, including ridge rolls, valleys, gutters and so on, are to be of 28-gauge full weight.

4. All eaves trough, conductor pipe, shoes, mitres and all accessories, including gutters, valleys, ridge rolls and so on, when made of copper, to be not lighter than 16 ounces.

The 2″ pipe should be used only for small roofs such as porches, etc., where there is a small amount of water to be carried away.

The table below gives the nominal and actual sizes of the square corrugated pipe listed above:

Nominal Size.....	2″	3″	4″	5″
Actual Size.......	1¾″ x 2¼″	2⅜″ x 3¼″	2¾″ x 4¼″	3¾″ x 5″

Snow Guards

Snow guards are a necessary accessory to most slate roofs in sections of the country where the snowfall is sufficient to accumulate masses of snow and ice which are

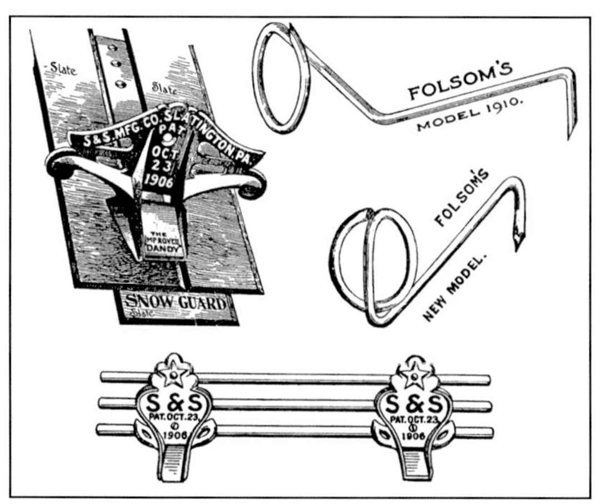

Figure 46. Snow Guards

liable to slide from the roof. On the slate roofs of the Harriman estate, along the Hudson, 35,000 copper wire snow guards were used.

The appearance of many slate roofs in such localities has been marred by streaks of rust stains from such guards, or snow brakes, as they are sometimes called.

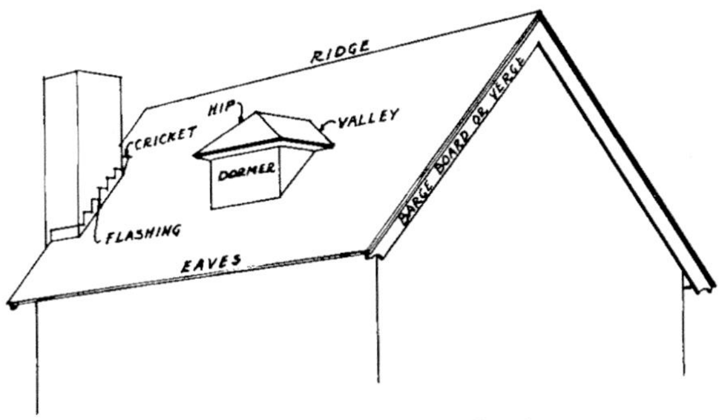

Figure 47. Parts of a Roof

Years ago, according to one manufacturer, they marketed 90 black-painted guards to every 10 galvanized. But now, since research shows wisdom of proper installations and better construction is demanded, this manufacturer says he sells 75 galvanized to every 25 black-painted.

While the "hot dip" process of galvanizing is the best, experts have still to be convinced that any galvanizing is perfect or rust proof. Hence, for first-class jobs only non-ferrous metals should be used for snow guards or any slate roof accessories.

Snow guards should be placed in manner prescribed by their manufacturers on all slate roof surfaces above doorways, porches, sidewalks, playgrounds or places where people are liable to pass or gather. They are essential to prevent masses of snow and ice from falling and as a protection to lower roof surfaces and gutters from such sliding masses. They are absolutely indispensable to retain snow on roofs when water from roof is collected for cistern or household purposes.

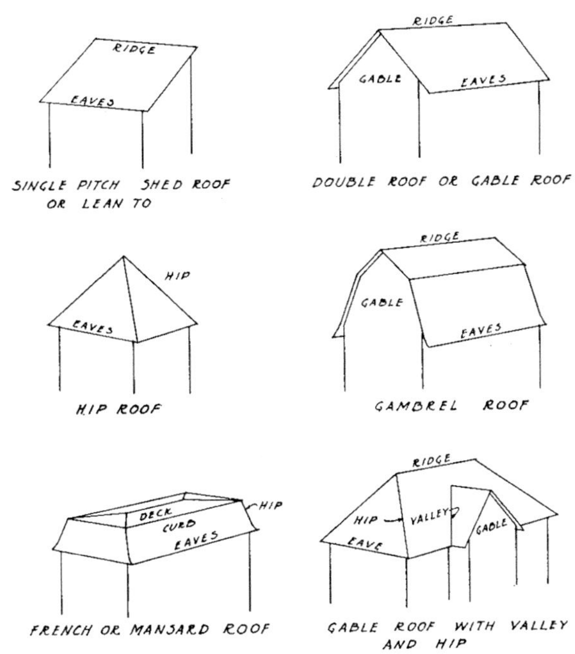

Figure 48. Types of Roofs

Snow guards are made in various forms, some of which are illustrated in Figure 46. Each type requires different methods of application and may be obtained from slate distributors or quarriers of roofing slate or from manufacturers. The Association can furnish a list of the latter concerns if desired. Some types may easily be placed on old roofs.

Lightning Rods

Where lightning rods are desired or used it is important at all times that they be properly grounded. All connections must be electrically perfect; that is, they must have very low contact resistance. Tips of rods should be silver, gold or platinum plated and all sharp bends in the conductor cable avoided.

All extensive masses of metal such as water or gas pipe systems or cast iron soil pipe should be connected to the ground by soldered connections of not less than No. 10 B. & S. wire, preferably stranded.

It is generally assumed that a lightning rod protects the area and any structures included within a 45° angle cone whose apex is the tip of the rod itself.

Types of Roofs and Their Parts

To the majority of the users of this Manual, the types of roofs and their parts will need no explanation but for the benefit of those not familiar with the names and technical descriptions used herein, reference to the accompanying illustrations will provide the lay reader with the elementary information essential to a full understanding of the parts and types of roofs mentioned.

Piling Slate

In piling slate the important factors are the foundations, starting the piles, arrangement of piles and individual slates and the separations of the tiers.

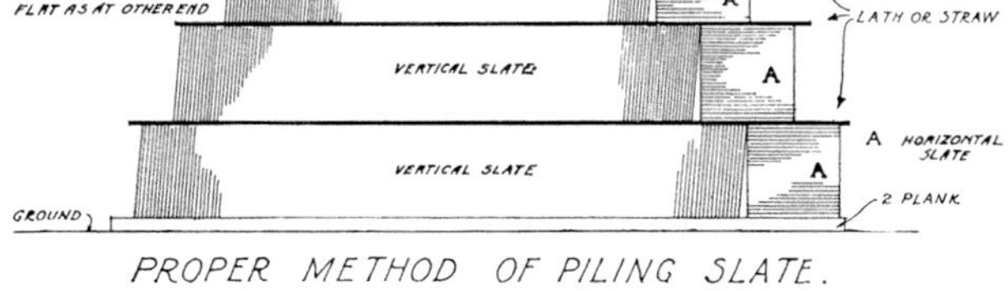

Figure 49

The earth foundation upon which the slate are to be piled should be level, dry and solid. A layer of two-inch plank will keep the slate off the ground, help to distribute the load and assist in maintaining straight even piles. The tiers of slate cannot be kept level if the foundation is not even or free from settlement.

The first tier should be started by laying one pile of slates flat to a height equaling the width of slate being piled; *i.e.*, for 20″ x 12″ slate the flat pile is 12″ high.

The following slates of the first tier are placed in an upright position on edge lengthwise, and should be kept as straight and vertical as possible. The bottoms of each handful should be tight against the bottom of the preceding slates. In this way the top is maintained straight and level.

After the first tier has been laid to the desired or a convenient length, lay a double row of wooden lath lengthwise over the top of the first tier. Place the lath 1″ from the outside edges of the slate tiers and interlap each lath one or two inches. A liberal quantity of straw may be used as a substitute for the lath.

Rest the flat or starter of the second tier one half on the first tier starter and one half on the upright slates. This will help to prevent the piles from overturning and "slumping" down obliquely. Keeping the following slates as nearly perpendicular as possible is especially important in the first two tiers.

Slates up to and including 20″ x 11″ may be safely piled up to 6 tiers high. Slates of a larger size should never be piled more than 4 tiers high. Closely piled, 100 commercial standard slates average 20″ to 24″.

When the slates are stored in an open yard, cover the piles with overlapping boards or use tar paper weighted down. Adequate protection prevents the slates from being frozen together. While slates are of ample strength when used in their proper place, reasonable care should be used in the handling of the material.

Estimating Slate Quantities and Costs

Architects and others interested should confer with local roofing contractors and through them obtain the "per square" or approximate total price for a particular type of roof, or have them estimate the quantities required and furnish a definite price for each individual roof. Producers are always glad to co-operate with roofing contractors to furnish architects a layout or recommendations for slate suitable for any design or desired effect and to furnish roofing contractors estimates so they can give architects or owners comparative cost data to assist them to arrive at amount to spend for slate roof.

A mistake frequently made is that of measuring the roof surface and assuming it as the exact or very close approximate quantity required, then multiplying it by a price quoted in the price-list of one of the slate companies and perhaps adding an assumed sum for labor and contractors' profit.

In preparing estimates for a slate roof, every roofing contractor has his own method of making compensating allowances for waste, breakage, projections through roofs, dormers, hips, ridges, valleys and other factors occasionally entering into the question of quantity and labor. These are usually based upon experience with his own regular labor, local conditions or practice, and possibly other items peculiar to a locality. The following suggestions are therefore offered mainly for architects and others not actively engaged in selling and laying slate roofs, as a guide to many factors which should to be taken into consideration. For rough estimates a good roof costs from 6 to 8 per cent of total cost average building or home.

From the time work is first started at the quarry until the material is laid on the building, there are certain costs which must be taken into consideration. These may be listed as follows:

1. Cost of slate (punched) on cars at the quarry.
2. Freight from quarry to destination.
3. Loading and hauling to storage yard.
4. Unloading, piling and waste at storage yard.
5. Loading and hauling to job.
6. Unloading and piling at job.
7. Placing on roof and laying.
   (a) Roofing felt. (b) Elastic Cement. (c) Nails.
   (d) Snow guards, or Snow Rails. (e) Sheet Metal.
   (f) Labor, including compensation insurance.
   (g) Waste in handling, cutting and fitting.
8. Contractor's overhead on organization and equipment.
9. Cost of guarantee or Bond.
10. Contractor's profit.

Items 3 and 4 are omitted when material is hauled direct to the job and unloaded. Freight to any locality from a quarry remains a fixed charge based upon either carload or less than carload lots. Less than carload lots carry a freight charge about double that of carload quantities. The question of loading, hauling, unloading and laying depends upon local labor costs, how fast the men work and nature of the contractor's equipment. Some roofing contractors have slate punching machines and buy their slate unpunched, punching same at the job or in the yard on idle days or during inclement weather.

While it may seem a comparatively simple problem to estimate the net quantity, it is not so easy to allow for the additional material required for slate around chimneys, dormers, hips, valleys, etc. These allowances depend largely upon the judgment and experience of the estimator and the roof design.

Method Suggested for Estimating

1. Obtain the net area of the roof in square feet, adding 6″ to rafter length to allow for waste of normal roof.
2. Deduct one-half of the area of chimneys and dormers if over 20 sq. ft. and less than 80 sq. ft. Make no deduction if less than 20 sq. ft., and deduct 20 sq. ft. less than actual area if more than 80 sq. ft.
3. Include areas of dormer roofs, sides of dormers if slated, slate saddles, or other places where slate is used in addition to the main roof area. Include overhanging parts of dormers, etc.
4. Add 1 sq. ft. for each lineal foot of hips and valleys, for loss in cutting and fitting.
5. Allow from 2% to 15% additional slate, depending upon the extent to which the roof is intersected by other roofs, dormers, walls, other contingencies, etc.
6. Divide the total of the above by 100, which will give the number of "squares" of roofing required.

It should be noted that slate is always sold at the quarry on the basis of the quantity required to cover "100 sq. ft." or a "square" of roof when slate is laid with a 3″ head lap. If the roof is flat or other than 3″ lap is used, the quantity must be corrected to the equivalent amount required as though the 3″ lap was used. The following information should be given local roofing contractor when asking for a price:

1. Kind and color of slate.
2. Size of slate desired, stating length and "all one width" or random width.
3. Thickness, as "commercial standard," ¼″, ⅜″, etc.
4. Type of roof, as standard, textural, graduated or flat.
5. Kind of nails, as zinc clad, zinc, "yellow metal," copper clad or copper.
6. Kind of valleys and flashings.
7. If hip or gable roof.
8. Kind of snow guards, as galvanized, yellow metal or copper.
9. If snow rails, size of pipe and number of rows of pipe.
10. Location of job; if in city or vicinity, or out of city.
11. When job is to be finished.

Prices

Many requests have been received from architects and builders for the inclusion in this book of a definite list of prices for the various grades and colors of roofing slates. The compilers regret that it is impracticable to do this, for several reasons. The Association recommends, therefore, that when an architect or builder desires the price of any particular grade and color of slate, he consult his local slate roofing contractor for estimates for the slate in place on the roof. Such estimates will then include the many factors listed in the section on "Estimating," all of which must be given consideration. This can best be given by responsible contractors who are familiar with local conditions and costs.

Advantages of Slate

During the preparation of this book many requests to include in it a list of the advantages of slate in direct comparison with other roofing materials were received from architects, contractors, and especially from prospective home builders.

The policy of the National Slate Association is, and always has been, never to point out the faults or weaknesses of a competitive or a substitute material for any of the uses of slate. Such procedure conforms with sound business ethics, and, in line with this policy, the editors merely list in convenient form the outstanding characteristics which make slate such a valuable roofing material. Each of these qualities will be found discussed elsewhere in this book.

- Natural Stone
- Non-combustible—Fireproof
- Waterproof
- Permanent
- Wide range of effects possible
  - Appearance
  - Color
  - Thickness
  - Surface Texture
  - Roof Texture
- Little or no maintenance costs
- Resists climatic changes
- Requires no other material to preserve it
- Reduces insurance premiums
- High salvage value
- Increases property values

Investigation has shown that the cost of the roof of any structure ought to bear a certain definite relation to the cost of the entire building, not only from the standpoint of protection, but also from appearance. Every permanent building should be roofed with a material which will give it lasting protection, and at the same time be in character with the house itself and its surroundings. Because it does possess these qualities, for centuries slate has been the criterion by which other roof coverings have been judged. A slate roof has that indefinable "something" which distinguishes any object of real value and completes the picture of a well constructed building or home.

GENERAL NOTES—ALL SPECIFICATIONS

(a) The following specifications for "Standard," "Textural" and "Graduated" slate roofs give in detail the procedure to be followed in the laying of each type of roof on any type of structure from the smallest bungalow to the largest mansion.

(b) A "Short Form" of each of these specifications is given just ahead of the "Basic Form" and is intended to be written into the architect's specification. The "Short Form" refers to the "Basic Form" of the National Slate Association as printed herein and carries with it all the provisions contained in the "Basic Form." In each "Short Form" the architect should fill in the color and size of the slate desired; otherwise it is complete as printed.

(c) It will be noted that the "Sheet Metal Work" is a separate specification. The architect or owner can use it as a separate specification if this work is the subject of a separate contract or add it to the slate roofing specification if it is to be part of the slate roofing contractor's work as is the custom in certain localities.

(d) In each specification it is assumed that sheathing boards, crickets, cant strips or other under-roof surface will be specified elsewhere under the proper trade.

(e) The marginal notes are informative and indicate optional changes which may be made in the specifications.

(f) These specifications do not in any case include General Conditions or items relative to administrative matters such as usually form the first part of an architect's specification. Such items include fire, compensation, liability or other insurance and the use of hoists, water, telephone, watchman, temporary heat and light, and storage.

(g) The attention of the architect and owner is called to this fact in order that they may be fully acquainted with all conditions and advise the roofing and sheet metal contractor to arrange for the proper disposition of such items with the owner when the contract is direct or with the general contractor when his status is that of a sub-contractor.

(h) The National Slate Association recommends the use of the Standard Documents of the American Institute of Architects and especially the "Standard Form of Sub-contract" which is a form of agreement between the General Contractor and the Sub-contractor. It calls attention, however, to paragraphs f and g above as these subjects are only covered in a general way in the "Standard Form."

STANDARD SPECIFICATIONS
FOR
SHEET METAL WORK
IN CONNECTION WITH A SLATE ROOF

NOTES

1. This specification used as here given will provide for sheet metal work in connection with a slate roof as follows:
   Paper—Rosin sized or asbestos felt, 6 lbs per sq. (unless already covered for slate)
   Nails—Copper. Flashings—16 oz. copper.
   Valleys—Open.
2. The following must be written into the specification, if required:
   Decks, Cornices, etc.
   Gutters and Downspouts.
3. See also General Notes—All Specifications, page 48.

SHORT FORM SPECIFICATION

1. Sheet Metal Work

All materials and labor in connection with all sheet metal work shall be furnished and performed in strict compliance with the recommended practice and Standard Specification "M" for Sheet Metal Work, 1925, of the National Slate Association, 791 Drexel Building, Philadelphia, Penna.

2. Decks, Cornices, Etc. *If required specify materials and methods.*

3. Gutters and Conductor Pipes *If required specify materials and methods.*

STANDARD SPECIFICATION "M"

(Covering in detail Flashings, etc., in connection with a Slate Roof)

1. General and Guarantee

(a) The General conditions of the contract are hereby made a part of the contract and this contractor shall examine these General Conditions and thoroughly acquaint himself with all the requirements therein contained.

(b) This contractor shall furnish all materials and labor in accordance with the drawings and these specifications.

(c) This contractor shall carefully examine all surfaces prepared for flashings, etc., by other trades, shall point out all defects to the proper authority, and shall see that the necessary corrections are made before proceeding with his work.

This contractor shall arrange his work so as to co-operate at all times with othér trades and prevent delay or damage to other work.

(d) During construction care shall be taken to prevent damage to flashings in place by walking or placing heavy materials on them. As soon as soldering is done and flashings are completed, the work shall be thoroughly cleaned. Toward completion, all damaged work shall be repaired, all stains and débris shall be removed, and sheet metal work shall be left in perfect condition.

(e) The contractor shall furnish his own scaffold or rigging, or arrange with the general contractor for the use of scaffolds furnished by others.

(f) A written guarantee shall be furnished that the materials used are in strict accordance with these specifications, and that any and all repairs required on the roof due to defective materials or workmanship furnished under this contract shall be made without cost to the owner, for a period of one year.

Image from 1933 cigarette advertisement.

2. Preparation of Surfaces

All surfaces to receive flashings shall be made smooth and even, and all nail heads shall be set.

3. Building Paper

All surfaces to be covered with copper shall be covered first with rosin-sized or asbestos-felt paper weighing not less than 6 pounds per 100 square feet. Paper shall lap 2" and be nailed with flat-head copper nails. If surfaces have already been covered with paper or felt by the roofing contractor this paper may be omitted.

4. Nails and Fastenings

All nails, rivets and similar fastenings, if any, used throughout the work shall be of best grade hard copper or brass. Nails shall be wire nails not less than No. 12 gauge and not less than 7/8" long.

5. Copper

All flashings shall be 16 ounce soft copper, rolled from copper conforming to the standard specification of the American Society for Testing Materials. The edges of all sheets to be soldered shall be tinned 1½" on both sides.

If other than copper is to be used, specify material, gauge, weight, painting and method of laying.

6. Solder and Rosin

All solder shall be of the best grade, equal to Specification B-32-21 of the American Society for Testing Materials, and shall be composed of one-half pig lead and one-half block tin (new metals). Rosin shall be used as a flux.

7. Exposed Edges

The exposed edge of all flashings shall be folded under ½″, in in such manner as to conceal them from view.

8. Flashings—Where Required

All intersections of roofs with vertical surfaces of every kind and all openings in roof surfaces, shall be flashed with copper. The method of flashing, except as otherwise shown or specified, shall be base and counter, or cap, flashing.

9. Base Flashings

(a) Unless otherwise specified or shown on the drawings, base flashings shall be, in general, at least 4″ high. They shall project at least 4″ out on to the roof. Flashings shall be full 96″ in length. On sloping roofs they shall lap longitudinally at least 3″. On flat roofs the joints shall be flat-locked and soldered.

(b) Against stucco-coated walls, metal lath shall lap outside the flashing so that the stucco shall finish over the flashing.

10. Cap Flashings or Counterflashings

Cap flashings shall turn down over base flashings not less than 4″. They shall be secured to vertical surfaces, as follows:

Wood Work: They shall extend up under exterior coverings not less than 4″ above the roof, and shall be nailed along the top edge about every 8″.

Mason Work: They shall extend into joints of masonry walls 4″ and have the inner edge turned back on itself ½″. The sheets shall be bent to the required shapes, and built in with the mason work. No cutting out of joints for setting flashings will be allowed.

Reglets: Where indicated on the drawings or where specified, flashings shall finish in reglets in the masonry cut by others where located by this contractor.

The flashing shall be turned into the reglet the full depth and shall be turned back to form a hook.

After the flashing is in place the reglet shall be filled and caulked, using molten lead on flat surfaces, and lead wool on vertical surfaces.

After caulking the reglet shall be made smooth by filling with elastic cement.

Stucco on Wood: When used with stucco-covered wood-frame walls, cap flashings shall be formed over a ⅞″ base board and extend up the wall at least 2″ above the base board, and be nailed at the top edge with nails about 12″ apart. Metal lath shall be placed over the flashing and the stucco should be finished against the base board.

Stucco on Masonry: They shall be built into the masonry as the work progresses and shall project out from the wall as required and turn down over the base flashing. The stucco shall finish against the cap flashing.

Concrete Walls: They shall be set in the forms before the concrete is poured. They shall extend into the wall at least 2″ and shall have the inner edges turned back ½″.

From the Slate Roof Bible 3rd Edition (2016)

Specification writer should include here only those paragraphs which apply to the work and make suitable provision in mason and carpenter specification for preliminary work.

From the Slate Roof Bible 3rd Edition (2016)

Make provision in proper specification for baseboard, metal lath, etc.

11. Step Flashings

Step flashings shall be used where vertical surfaces occur in connection with slopes. They shall be formed of separate pieces built into the masonry as specified for cap flashings in masonry. Steps shall lap generally 3″, but in no case less than 2″, and shall not be soldered. Lap joints shall be vertical.

12. Vent Flashings

All pipes passing through roofs shall be flashed and counter-flashed. Base flashings shall extend out on the roof not less than 6″. They shall be of sufficient length to cover the slate course next below the pipe and to extend up under the slate course above as far as possible without puncture by nails.

Patented vent-flashing devices may be used, subject to the approval of the architect. They shall be made of 16 ounce copper, shall be the product of a recognized manufacturer, and shall be installed according to manufacturer's directions.

If closed valleys are desired, specify as follows:

Flashing pieces, for closed valley shall be of sufficient length to extend 2 inches above the top of slate roofing piece and lap the flashing piece below 3 inches, and of width sufficient to extend up the sides of the valley far enough to make the valley 8 inches deep.

They shall be placed with the slate so that all pieces are separated by a course of slate. Pieces shall be set so as to lap at least 3 inches and to be entirely concealed by the slates. They shall be fastened by nails at the top edge only.

13. Open Valley Flashings

(a) Open valleys shall be not less than 4″ wide. The proper width shall be determined by the following rule:

Starting at the top with a width of 4″, increase the width one inch for every 8 feet of length of the valley. Flashing pieces shall be full length sheets and of sufficient width to cover the open portion of the valley and extend up under the slate not less than 4″ on each side.

A small inequality in the areas drained would not require this.

(b) Where two valleys of unequal size come together, or where the areas drained by the valley are unequal, there shall be placed in the valley a "crimp" angle or tee not less than 1″ high. This "crimp" may be formed in the valley sheet before placing, or it may be made of a separate piece soldered to the valley sheet.

Specify framing and blocking for crickets under carpentry.

14. Crickets or Saddle Flashing

Crickets or saddles formed back of all vertical surfaces, such as chimneys, etc., breaking through sloping roofs, shall be covered with copper. The flashing of these crickets shall be made of part of the flashing along the sides of the chimney, etc.

Make provision in mason specifications for building in these flashings, and the cutting of any necessary reglets.

15. Other Work

This contractor shall furnish all flashing pieces to other trades such as the Mason, Carpenter and Roofing Contractor for building in with that particular work and shall co-operate with these contractors in doing their work.

16. Decks, Cornices, Etc.

If required specify materials and methods.

17. Gutters and Down Spouts

If required specify materials and methods.

Standard Specification "M"—Sheet Metal Work, 1925

Pennsylvania Slate Institute
PEN ARGYL - PENNSYLVANIA
CERTIFICATE GUARANTEE
This Certifies THAT THE SLATE LOADED IN CAR NO. ________ ________
is Pennsylvania ________ Slate
of the following Pennsylvania Slate Institute standard of quality:
Of a modulus of rupture of over 7,000 pounds per square inch. (A.S.T.M. tentative standard test).
Of a porosity of less than ½ of one per cent (A.S.T.M. tentative standard test).
Free from longitudinal curvature to the extent of less than ¼″ variation in 16″ slates or ⅛″ in 24″ slates.
Of a thickness to average 22″ to 24″ to the hundred slate.
Attested by ________
Certificate No. ________ Date ________

From the Slate Roof Bible 3rd Edition (2016)

STANDARD SLATE ROOFS

The "Standard" slate roof has been developed to provide a serviceable, permanent roof at a minimum cost. Any changes which have been made in the usual specification do not in any way affect those valuable features which are to be found only in a slate roof but merely eliminate so far as possible every factor which might increase the cost of slating.

As has previously been explained, the nature of slate stone is such that every piece cannot be split to an exact thickness and in order to eliminate a selection of the material with its resultant rehandling and calipering of each piece, a slight variation in the thickness of the slates is permitted. "Commercial Standard" slates are the quarry run of production and may show tolerable variations above or below 3/16".

A second change is found in the note opposite paragraph 3, page 53, of the Specification. The Architect or Owner will find it to his advantage to confer with the roofing contractors in his locality regarding the size (length and width) or sizes of the slate economically available. The number of sizes shown on page 12 is due to the nature of the material which makes it necessary to cut the blocks into a large number of different sized pieces to secure the fullest utilization of the slabs as they are taken from the quarry.

It will therefore be readily appreciated that certain sizes may be more easily obtained than others and that to insist on slates of a definite dimension may not only add to the cost of the roof but may delay the work owing to difficulty in locating a quarry having that particular size in stock. On roofs of large area sufficient economy can often be effected by a proper selection of sizes to make a preliminary investigation of value. The smaller sizes ranging from 10" x 6" to 14" x 10" are usually more plentiful than the larger sizes and when laid with a narrow exposure will be found to produce a most interesting and artistic effect similar to that secured in the same way on the wonderful chateaux of France.

STANDARD SPECIFICATION
FOR A
STANDARD SLATE ROOF

NOTES

1. This specification used as here given will provide for a complete slate roof as follows:

   Felt—Saturated Asphalt, 14 lbs. per sq.
   Flashings—16 oz. copper
   Valleys—Open
   Hips—Saddle
   Cement—Elastic waterproof
   Ridges—Saddle
   Nails—Copper
   Slate—Commercial standard (approx. 3/16")

2. The following must be written into the specification under paragraph:

   3-b. Slate—size
   3-c. Slate—color

3. It should be noted that metal work is not covered by this specification, as separate Standard Specifications for Sheet Metal Work are given on page 48. The specifications for Slate Roofing and sheet metal work are so arranged, however, that they may be handled separately or together to conform to variation in custom in different sections of the country.

4. See also General Notes—All Specifications, Page 48.

SHORT FORM SPECIFICATION

NOTE: In using the Short Form if any optional paragraphs are desired so state.

1. Slate Roofing

(a) Furnish all materials and labor and cover with commercial standard slate all portions of the building shown, marked or indicated for Slate, and form all slate hips and ridges in accordance with Specification S for a Standard Slate Roof, 1925, of the National Slate Association, 791 Drexel Building, Philadelphia, Pa.

(Continued on Page 53)

For standards and optional materials and methods, see the following paragraphs in the complete form.

Paragraph 1—Guarantee
Paragraph 2—Roofing Felt
Paragraph 3—Slate
Paragraph 4—Hips
Paragraph 5—Ridges
Paragraph 6—Valleys
Paragraph 7—Elastic Cement
Paragraph 8—Nails
Paragraph 9—Slating

(b) The size of the slate shall be ______________.
(c) The color of the slate shall be ______________.

Paragraph 3-b—Size of Slate
Paragraph 3-c—Color of Slate

2. **Sheet Metal Work**

______________________________.
______________________________.

If part of the slate roofing contractor's work, quote here the Short Form for Sheet Metal Work on page 49.

STANDARD SPECIFICATION "S"

(Covering in detail all materials and labor in connection with a Standard Slate Roof)

1. **General and Guarantee**

(a) The General Conditions of the contract are hereby made a part of the contract and this contractor shall examine these General Conditions and thoroughly acquaint himself with all the requirements therein contained.

(b) This contractor shall furnish all materials and labor in accordance with the drawings and these specifications.

(c) This contractor shall inspect all surfaces prepared for slating by other trades, point out to the proper authority all defects, and shall not proceed with the laying of felt, flashings or slate until the necessary corrections have been made.

(d) Roofing shall be applied by workmen experienced in the applying of slate.

(e) The roofing contractor shall furnish his own scaffold or rigging, or arrange with the general contractor for the use of scaffolds furnished by others.

(f) A written guarantee shall be furnished that the materials used are in strict accordance with these specifications, and that any and all repairs required on the roof due to defective materials or workmanship furnished under this contract shall be made without cost to the owner, for a period of one year.

2. **Roofing Felt**

(a) On all boarding to be covered with slate, furnish and lay asphalt saturated rag felt, not less in weight than that commercially known as "14 pound" felt. In this weight, per 100 square feet, a tolerance of one pound plus or minus will be allowed.

(b) Felt shall be laid in horizontal layers with joints lapped toward eaves and at ends at least 2″ and well secured along laps and at ends as necessary to properly hold the felt in place and protect the structure until covered by the slate. All felt shall be preserved unbroken, tight and whole.

(c) The felt shall lap over all hips and ridges.

(d) Felt shall be lapped 2″ over the metal of any valleys or built-in gutters.

3. **Slate**

(a) Slate shall be of commercial standard quality and thickness.

Photos adapted from the Slate Roof Bible, 3rd Edition (2016)

(b) Size of slate shall be ______________, or random widths of ______________ length(s).

(c) Color of slate shall be ______________.

For list of slate colors see page 7.
If weathering slate are desired, so state, otherwise unfading colors will be furnished.
If a particular quarry product is desired, so state.

For additional data regarding roofing felt, see page 23.

Commercial standard slate is approximately 3/16″ in thickness. If any other thickness is desired, so state and omit "Commercial."

For table of standard sizes, see page 12.
A definite size must be stated to place all estimates on an equal basis.
To obtain a roof at a minimum cost, confer with local roofers as to the size or sizes economically available. Standard roofs are usually laid with slate of one size on any one area. If random widths are desired, so state.

(Continued on Page 54)

(d) All slate shall be hard, dense, sound rock, machine punched for two nails each. All exposed corners shall be practically full. No broken corners on covered ends which sacrifice nailing strength or the laying of a watertight roof will be allowed. No broken or cracked slates shall be used.

"Machine punched" does not exclude hand punching by means of the punch and mawl or hand punching with the hammer, as the slate is laid, to properly locate holes for fitting hips, etc., or obtaining suitable nailing to the roof boards.

4. Hips

(a) All hips shall be laid to form "Saddle" Hips without metal underneath.

If any other type of hip is desired, so state.

5. Ridges

(a) All ridges shall be laid to form "Saddle" Ridges. The nails of the combing slate shall pass through the joints of the slates below.

If combed ridges, ridge rolls or cresting are desired, so state.

6. Valleys

(a) All valleys shall be laid to form "Open" Valleys.

If "Closed" Valleys are desired, so state.

7. Elastic Cement

(a) Cement shall be an approved brand of waterproof elastic slaters' cement colored to match as nearly as possible the general color of the slate.

8. Nails

(a) All slate shall be fastened with two large-head slaters' solid copper nails. Use 3d (1¼″) nails for slates 18″ or less in length, 4d (1½″) for 20″ or longer, and 6d (2″) for slates on hips and ridges.

9. Slating

(a) The entire surface of all main and porch roofs, the roofs and sides of any dormer windows, if shown, and all other surfaces so indicated on the drawings, shall be covered with slate in a proper and watertight manner.

(b) The slate shall project 2″ at the eaves and 1″ at all gable ends, and shall be laid in horizontal courses with the standard 3″ headlap, and each course shall break joints with the preceding one. Slates at the eaves or cornice line shall be doubled and canted ¼″ by a wooden cant strip.

(c) Slates overlapping sheet metal work shall have the nails so placed as to avoid puncturing the sheet metal. Exposed nails shall be permissible only in top courses where unavoidable.

(d) Neatly fit slate around any pipes, ventilators, etc.

(e) Nails shall not be driven in so far as to produce a strain on the slate.

(f) Cover all exposed nail heads with elastic cement. Hip slates and ridge slates shall be laid in elastic cement spread thickly over unexposed surface of under courses of slate, nailed securely in place and pointed with elastic cement.

(g) Build in and place all flashing pieces furnished by the sheet metal contractor and co-operate with him in doing the work of flashing.

(h) On completion all slate must be sound, whole and clean, and the roof shall be left in every respect tight and a neat example of workmanship.

Vertical surfaces may be laid with 1″ headlap.

Specify cant strips of proper thickness under Carpentry.

Where cant strips occur above gutters, they should be placed before the metal.

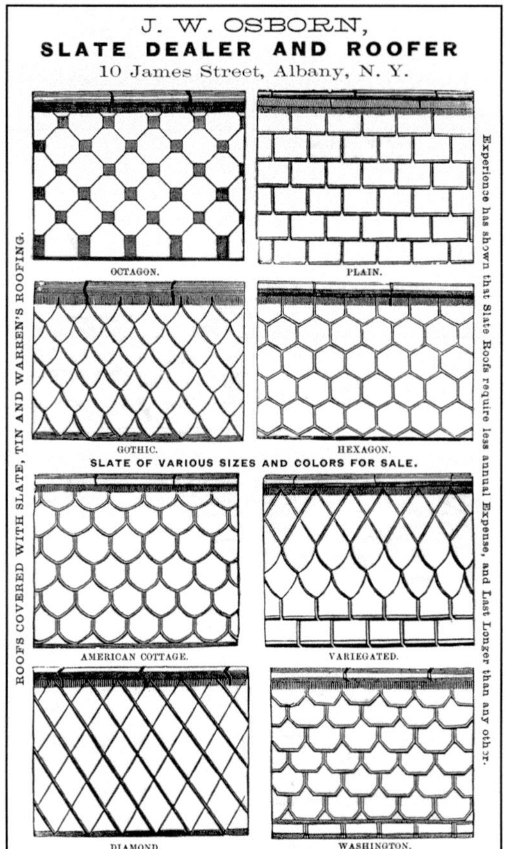

1864 Advertisement

From the Slate Roof Bible, 3rd Edition (2016)

TEXTURAL SLATE ROOFS

The term "Textural" is applied to slate roofs in this publication for the first time, so far as is known. It has been here coined to designate those charming roofs of slate which cannot be classified as strictly Standard or Graduated but which range between the two and produce a distinctive and altogether individual result.

The Textural roof presents most fascinating possibilities to the designer, for it may be varied in almost hundreds of ways to conform and harmonize with the structure of which it is to become a part. It may be made nearly as chaste as the Standard or it may be almost as elaborate as the Graduated. For example, the slates may be uniform in length but vary in thickness, or the thickness may be kept uniform and the length (and exposure) varied. The width may vary or be uniform. Very rough surfaced slates may be introduced into an otherwise almost smooth roof. The slate may be laid with varying exposure or the butts may be chipped and broken at irregular angles. When the thickness is varied the usual range is from 3/16″ to ⅜″ but if a heavier and more irregular effect is desired, this may be accomplished by incorporating certain percentages of thicker slates.

These present only a few of the opportunities for interesting treatment. In addition, the color of the slates may be used with splendid effect by mixing slates of different tones, but the tone value or predominating color effect desired should always be specified. When color is used, accidental or "freak" slates may be introduced throughout the roof, adding greatly to the picturesque effect of the ensemble.

The cost of a Textural roof will depend entirely upon the effects desired, and although usually slightly greater than the Standard roof, will not equal in either material or labor the cost of a Graduated roof.

The various details of construction, sheathing, ridges, hips, valleys, flashings, etc., shown in connection with the Standard slate roof apply also to the Textural roof. The construction may require additional collar beams, supports or slightly heavier rafters than for a Standard roof, depending upon local conditions and construction as well as the weight of the slate selected.

STANDARD SPECIFICATIONS
FOR A
TEXTURAL SLATE ROOF

NOTES

1. This specification used as here given will provide for a complete slate roof as follows:

Felt—Saturated Asphalt commercially known as 30 lbs. per sq.
Flashings—16 oz. copper.
Valleys—Closed.
Cement—Elastic Waterproof.
Nails—Copper.
Slate—3/16″ to ⅜″, random width, approved lengths.
Hips—Saddle.
Ridges—Saddle.

2. The following must be written into the specification under paragraph:

3-c Slate—Predominating color.

3. It should be noted that metal work is not covered by this specification, as separate Standard Specifications for Sheet Metal Work are given on page 48. The specifications for Slate Roofing and sheet metal work are so arranged, however, that they may be handled separately or together to conform to variation in custom in different sections of the country.

4. See also General Notes—All Specifications, Page 48.

SHORT FORM SPECIFICATION

NOTE: In using the Short Form if any optional paragraphs are desired so state.

1. Slate Roofing

(a) Furnish all materials and labor and cover with slate all portions of the building shown, marked or indicated for slate, and form all slate hips and ridges in accordance with the Specification "T" for a Textural Slate Roof, 1925, of the National Slate Association, 791 Drexel Building, Philadelphia, Pa.

(Continued on Page 56)

For standards and optional materials and methods. see the following paragraphs in the complete form.

Paragraph 1—Guarantee
Paragraph 2—Felt
Paragraph 3—Slate
Paragraph 4—Hips
Paragraph 5—Ridges
Paragraph 6—Valleys
Paragraph 7—Elastic Cement
Paragraph 8—Nails
Paragraph 9—Slating

(b) The predominating color of the slate shall be ————.

Paragraph 3-b—Sizes of Slate
Paragraph 3-c—Color of Slate

2. **Sheet Metal Work**

————————————
————————————

If part of the slate roofing contractor's work, quote here the Short Form for Sheet Metal Work on page 49.

STANDARD SPECIFICATION "T"

(Covering in detail all materials and labor in connection with a Textural Slate Roof.)

1. **General and Guarantee**

(a) The General Conditions of the contract are hereby made a part of the contract and this contractor shall examine these General Conditions and thoroughly acquaint himself with all the requirements therein contained.

(b) This contractor shall furnish all material and labor in accordance with the drawings and these specifications.

(c) The roofing contractor shall furnish samples of the slate he proposes to use. If required, a layout shall be prepared by the producer and submitted to the Architect for approval.

(d) This contractor shall inspect all surfaces prepared for slating by other trades, point out to the proper authority all defects, and shall not proceed with the laying of felt, flashings or slate until the necessary corrections have been made.

(e) Roofing shall be applied by workmen experienced in the applying of slate.

(f) The roofing contractor shall furnish his own scaffold or rigging, or arrange with the general contractor for the use of scaffolds furnished by others.

(g) A written guarantee shall be furnished that the materials used are in strict accordance with these specifications and the samples submitted, and that any and all repairs required on the roof due to defective materials or workmanship furnished under this contract shall be made without cost to the owner, for a period of one year.

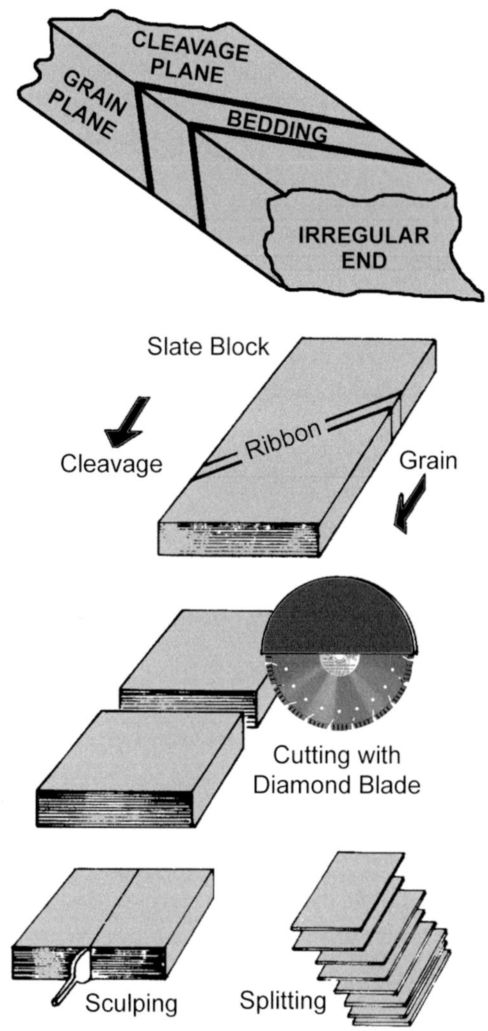

Illustration above adapted from National Park Service Preservation Brief #29 (1993)

2. **Roofing Felt**

(a) On all boarding to be covered with slate, furnish and lay asphalt saturated rag felt, not less in weight than that commercially known as "30 pound" felt.

(b) Felt shall be laid in horizontal layers with joints lapped toward eaves and at ends at least 2″ and well secured along laps and at ends as necessary to properly hold the felt in place and protect the structure until covered by the slate. All felt shall be preserved unbroken, tight and whole.

(c) The felt shall lap over all hips and ridges.

(d) Felt shall be lapped 2″ over the metal of any valleys or built-in gutters.

For additional data regarding roofing felt, see page 23.

3. **Slate**

(a) Slate shall be in combinations of thicknesses from 3/8″ to 3/16″ thick.

(b) Slates shall be of random widths and of such lengths as approved by the Architect.

If a percentage of thicker slates are desired, state the thickness and percentage of each.

If a layout is required, state—"in accordance with approved layout." If definite sizes must be furnished, so state. To obtain a roof at minimum cost, confer with local roofers as to sizes economically available.

(c) The predominating color of the slates shall be ————.

For list of slate colors see page 7.
Colors will be of the "weathering" class unless otherwise noted.

(d) All slate shall be hard, dense, sound rock, machine punched for two nails each. No cracked slate shall be used. All exposed corners shall be practically full. No broken corners on covered ends which sacrifice nailing strength or the laying of a watertight roof will be allowed. No broken or cracked slates shall be used.

"Machine punched" does not exclude hand punching by means of the punch and mawl, or hand punching with the hammer to properly locate holes for fitting hips, etc., or obtaining suitable nailing surface.

4. **Hips**

(a) All hips shall be laid to form "Saddle" Hips without metal underneath.

If any other type of hip is desired, so state.

5. **Ridges**

(a) All ridges shall be laid to form "Saddle" Ridges. The nails of the combing slate shall pass through the joints of the slates below.

If combed ridges, ridge rolls or cresting are desired, so state.

6. **Valleys**

(a) All valleys shall be laid to form "Closed" Valleys.

If open valleys are desired, so state. "Round" or "Canoe" valleys may be used if desired but will add to the cost of the roof and are more frequently used with Graduated Roofs. For data see page 22; for Specifications see Standard Specification for a Graduated Slate Roof, page 59.

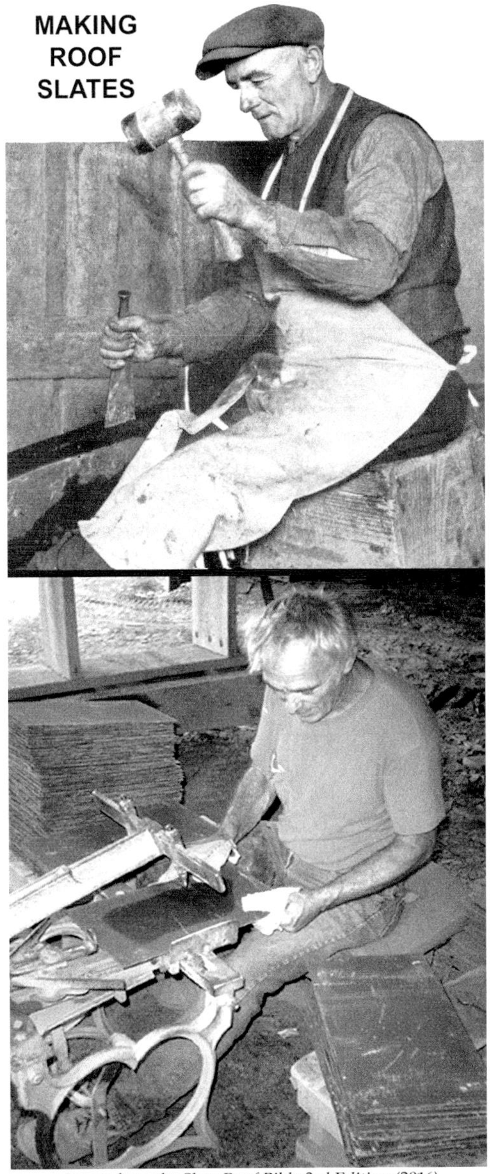

Image from the Slate Roof Bible 3rd Edition (2016)

7. **Elastic Cement**

(a) Cement shall be an approved brand of waterproof elastic slaters' cement colored to match as nearly as possible the general color of the slate.

8. **Nails**

(a) All slate shall be fastened with large-head slaters' copper nails of sufficient lengths to adequately penetrate roof boarding. Care shall be taken to avoid exposing the nails on cornice, soffits, overhanging eaves, etc.

9. **Slating**

(a) The entire surface of all main and porch roofs, the roofs and sides of any dormer windows, if shown, and all other surfaces so indicated on the drawings, shall be covered with slate in a proper and watertight manner.

(b) The slate shall project 2″ at the eaves and 1″ at all gable ends, and shall be laid in horizontal courses with the standard 3″ headlap and each course shall break joints with the preceding one. Slates at the eaves or cornice line shall be doubled and canted by a wooden cant strip.

(c) Slates overlapping sheet metal work shall have the nails so placed as to avoid puncturing the sheet metal. Exposed nails shall be permissible only in top courses where unavoidable.

(d) Neatly fit slate around any pipes, ventilators, etc.

(e) Nails shall not be driven in so far as to produce a strain on the slate.

(f) Cover all exposed nail heads with elastic cement. Hip slates and ridge slates shall be laid in elastic cement spread thickly over unexposed surface of under courses of slate, nailed securely in place and pointed with elastic cement.

(g) Build in and place all flashing pieces furnished by the sheet metal contractor and co-operate with him in doing the work of flashing.

(h) On completion all slate must be sound, whole and clean, and the roof shall be left in every respect tight and a neat example of workmanship.

Vertical surfaces may be laid with 1″ headlap. Specify cant strips of proper thickness under Carpentry. Where cant strips occur above gutters, they should be placed before the metal.

Standard Specification "T"—Textural Slate Roof, 1925.

GRADUATED SLATE ROOFS

The origin of the graduated roof forms an interesting bit of history. Their use in Europe dates back several centuries before the days of standardization, to the time when sufficient slate to cover the roof was extracted from the quarry by primitive methods without regard to definite size or established thicknesses. Doubtless the sizes made were those most convenient. Sent to the building, the slates were evidently sorted to some extent and hooked to roof lath by means of oak pegs. If an accident occurred to the roof, a messenger and cart would be dispatched to the nearest quarry for a "load of slate," no attention probably being paid to size or thickness, and the roof was patched and repaired as well as the material might allow. As a result, a careful study of the roofs of former years discloses the fact that the charm of such roofs may be traced to mass, slope, the mellowing effect of age and weather, and the irregular texture, as a result of roughness of the surface and variations in length, width and size, all being more or less random.

The graduated slate roof is the custom-made roof of the industry, and is therefore subject to individual characteristics and many variations to meet contingencies of design and conditions.

In designing a roof of this type, the method of graduating will depend upon the size of the roof and its span, the "scale" of the building upon which it is used, and the general effect desired. A properly designed graduated roof should show no decided or noticeable break between the various thicknesses, sizes, etc. An agreeable and harmonious result can only be obtained by care in design and laying and the selection of the correct colors, sizes and texture. The interpretation of the design of such roofs is so much a matter of experience and good judgment that certain producers who have made a specialty of this work should be consulted and their suggestions obtained as to how to obtain the architectural effects desired.

For most cases a maximum thickness of 1″ will be found entirely satisfactory and pleasing in effect. The approximate thicknesses always obtainable are 3/16″, 1/4″, 3/8″, 1/2″, 3/4″, 1″, 1 1/4″ and 1 1/2″. The maximum practical thickness which can be used is 2″, but this is rarely used except for unusual conditions.

Once the architectural effect desired has been determined upon, the length of the roof slope and the resulting number of courses will determine the rate of decrease in thickness or the number of courses of any one thickness which can be used. The occasional introduction of slates of varying thicknesses in the same course in some roofs is regarded as a desirable feature and affords another method of influencing the irregular character of the surface. In general, the large, thick slates used near the eaves will occur in fewer courses than the thinner slates toward the ridge.

In addition to the usual standard sizes, slates above 1/2″ thick are produced in lengths up to 30″. The graduations in lengths generally range from 24″ to 12″. The longest slates are used at the eaves and the length usually diminished with the thickness. Here, again, good design dictates the range of graduations in length for any particular roof. The variations in length will at once provide a graduation in exposure by using the standard 3″ lap. To illustrate, a suitable range of lengths might be:

Under Eave Course	3/8″ thick	14″ long	No exposure
First Course	3/4″ "	24″ "	10 1/2″ "
1 "	3/4″ "	24″ "	10 1/2″ "
2 Courses	1/2″ "	22″ "	9 1/2″ "
2 "	1/2″ "	20″ "	8 1/2″ "
2 "	3/8″ "	20″ "	8 1/2″ "
4 "	3/8″ "	18″ "	7 1/2″ "
5 "	1/4″ "	16″ "	6 1/2″ "
3 "	1/4″ "	14″ "	5 1/2″ "
3 "	3/16″ "	14″ "	5 1/2″ "
8 "	3/16″ "	12″ "	4 1/2″ "

Random widths should be used and so laid that the vertical joints of each course are broken and covered by the slate of the course above. The variation in widths also tends to add interest to the texture and prevents a mechanical effect. The effect can be further enhanced by mixing percentages of heavier or lighter slates in each course of a given thickness.

In splitting the thick slate, a slightly roughened surface is obtained and the edges are more or less broken. The usual tendency is for the thinner slates to split cleaner and with greater evenness of surface. As a result, the slate toward the ridge are smoother in surface texture and should be selected for their roughness or at least sufficient used to avoid a flat, mechanical appearance in contrast with the heavier slates near the eaves. This is especially important since the ridge slate, being farther from the eye, will naturally tend to appear smoother and more even than those lower down the roof.

In addition to the variations described above, the color possibilities should not be forgotten. With proper selection the roof can be gradually blended from light to dark. In general, a dark roof appears to better advantage than a light color. Great care should be used in blending light to dark so there will be no pronounced line of demarcation of different shades.

Slates for graduated roofs may be had in all the natural slate colors, and many interesting effects may be obtained by the use of harmonious and contrasting colors laid at random. If not too "spotty," the weathering and ageing will tie the colors together, blending into a unit of unrivaled character.

Heavy slates require the best of nailing. Any slates over ¾″ in thickness and 20″ in length should be machine punched for four nails each. The smaller and thinner slates may be punched for two holes. Use only the best grade of non-ferrous slaters' nails with large heads and heavy gauge shafts of sufficient length to secure ample penetration into the supporting roof surface.

A heavy grade of felt will provide a cushion for the slates and provide additional insulation value. It is recommended that felt weighing from 30 to 50 pounds per square be used. (See Roofing Felt, page 23.)

It is decidedly unwise to use any but the best of flashing materials and workmanship on so fine a roof.

The construction of slate hips, valleys and ridges has been fully covered under the laying of slate, pages 14 to 23, and a repetition of this description is unnecessary here.

Rafters and their supporting members should be carefully proportioned to the load which they must sustain. Sheathing boards of ⅞″ material are ordinarily sufficient. When unusually thick slates are used, 1⅛″ sheathing boards should be specified.

STANDARD SPECIFICATION
FOR A
GRADUATED SLATE ROOF

NOTES

1. This specification used as here given will provide for a complete slate roof as follows:

Felt—Saturated Asphalt, 30 lbs. per sq.
Flashings—16 oz. copper.
Hips—Saddle.
Slate—¾″ to ¼″ thick, 24″ to 12″ long.
Cement—Elastic Waterproof.
Ridges—Saddle.
Nails—Copper.

2. The following must be written into the specification under paragraph:

3-c Slate—Color.
8-a Valleys—Type.

3. It should be noted that metal work is not covered by this specification, as separate Standard Specifications for Sheet Metal Work are given on page 49. The specifications for Slate Roofing and sheet metal work are so arranged, however, that they may be handled separately or together to conform to variation in custom in different sections of the country.

4. See also General Notes—All Specifications, Page 48.

SHORT FORM SPECIFICATION

NOTE: In using the Short Form if optional paragraphs are desired, so state.

1. Slate Roofing

(a) Furnish all materials and labor and cover with slate all portions of the building shown, marked or indicated for slate; form all slate hips, ridges and valleys, in accordance with Specification "G" for a Graduated Slate Roof, 1925, of the National Slate Association, 791 Drexel Building, Philadelphia, Pa.

For standard and optional materials and methods, see the following paragraphs in the complete form.

Paragraph 1—Guarantee
Paragraph 1—Samples and Layout
Paragraph 2—Felt
Paragraph 3—Slate
(Note: If other sizes and thicknesses than those in specification "G" are desired, specify exception at end of this paragraph.)
Paragraph 4—Hips
Paragraph 5—Ridges
Paragraph 7—Elastic Cement
Paragraph 8—Nails
Paragraph 9—Slating

(Continued on Page 60)

(b) The slate shall be ——————.
(c) The valleys shall be ——————.

Paragraph 3-c—Color of Slate
Paragraph 6—Valleys

2. Sheet Metal Work

——————————
——————————

If part of slate roofing contractor's work, quote here the Short Form for Sheet Metal Work on page 49.

STANDARD SPECIFICATION "G"

(Covering in detail all materials and labor in connection with a Graduated Slate Roof.)

1. General and Guarantee

(a) The General Conditions of the contract are hereby made a part of the contract and this contractor shall examine these General Conditions and thoroughly acquaint himself with all the requirements therein contained.

(b) The roofing contractor shall furnish samples of the slate he proposes to use. A layout showing graduations, courses and color distribution shall be prepared by the producer and submitted to the Architect for approval.

(c) This contractor shall furnish all materials and labor in accordance with the drawings and these specifications.

(d) This contractor shall inspect all surfaces prepared for slating by other trades, point out to the proper authority all defects, and shall not proceed with the laying of felt, flashings or slate until the necessary corrections have been made.

(e) Roofing shall be applied by workmen experienced in the applying of slate.

(f) The roofing contractor shall furnish his own scaffold or rigging, or arrange with the general contractor for the use of scaffolds furnished by others.

(g) A written guarantee shall be furnished that the materials are in strict accordance with these specifications, and that any and all repairs required on the roof due to defective materials or workmanship furnished under this contract shall be made without cost to the owner, for a period of one year.

Image added 2020

2. Roofing Felt

(a) On all boarding to be covered with slate, furnish and lay asphalt saturated rag felt, not less in weight than that commercially known as "30 pound" felt.

For additional data regarding roofing felt, see page 23.

(b) Felt shall be laid in horizontal layers with joints lapped toward eaves and at ends at least 2″ and well secured along laps and at ends as necessary to properly hold the felt in place and protect the structure until covered by the slate. All felt shall be preserved unbroken, tight and whole.

(c) The felt shall lap over all hips and ridges.

(d) Felt shall be lapped 2″ over the metal of any valleys or built-in gutters.

3. Slate

(a) Slate shall be in combination of thicknesses from ¾″ to ¼″.
(b) Slates shall be random widths from 24″ to 12″ in length.
(c) Color of slate shall be ——————.

Any desired range may be used, inserting the desired thicknesses and lengths.

State color or color mixture, permanent or weathering, and percentages of each desired.

(d) All slate shall be hard, dense, sound rock, machine punched for two holes. Slates ¾″ and thicker, when 20″ or more in length, shall have four holes. All exposed corners shall be practically full and no broken corners on covered ends which sacrifice nailing strength or the laying of a watertight roof will be allowed. No broken or cracked slates shall be used.

"Machine punched" does not exclude hand punching by means of the punch and mawl or hand punching with the hammer, as the slate is laid, to properly locate holes for fitting hips, etc., or obtaining suitable nailing to the roof boards.

4. Hips

(a) All hips shall be laid to form "Saddle" Hips without metal underneath.

If any other type of hips is desired, so state.

5. Ridges

(a) All ridges shall be laid to form "Saddle" Ridges. The nails of the combing slate shall pass through the joints of the slates below.

If combed ridges, ridge rolls or cresting are desired, so state.

6. Valleys

(a) All valleys shall be ————————.

Insert necessary following paragraphs:
1—Laid to form "Open" Valleys.
2—Laid to form "Closed" Valleys.
3—Of slate laid to radius. The slate shall be tapered, shouldered, securely nailed and shall be 4" longer than the slates used in the corresponding courses on the roof and laid to allow the tail coursings to line up. Proper copper flashing under each course.

7. Elastic Cement

(a) Cement shall be an approved brand of waterproof elastic slaters' cement colored to match as nearly as possible the general color of the slate.

8. Nails

(a) All slate shall be fastened with large flat-head slaters' copper nails of sufficient length to adequately penetrate nailing surface. Use four nails for slates $\frac{3}{4}''$ and thicker when 20″ or more in length. Care should be taken to avoid exposing the nails on cornice, soffits, overhanging eaves, etc.

(b) Nails securing slate must in no instance be driven through flashing.

From the Slate Roof Bible 3rd Edition (2016)

9. Slating

(a) The entire surface of all main and porch roofs, the roofs and sides of any dormer windows, if shown, and all other surfaces so indicated on the drawings, shall be covered with slate in combinations of colors, sizes and thicknesses shown on approved layout drawing furnished by Slate Producer.

(b) The slate shall project at the eaves and at all gable ends as directed, and shall be laid in horizontal courses with the standard 3″ headlap, and each course shall break joints with the preceding one. Slates at the eaves or cornice line shall be doubled and canted so that the succeeding course will have flat contact.

Vertical surfaces may be laid with a head lap of 1".
Specify cant strips of proper thickness under Carpentry.
Where cant strips occur above gutters, they should be placed before the metal.
If other construction at eaves and gables is desired, this should be noted here.

(c) Slates overlapping sheet metal work shall have the nails so placed as to avoid puncturing the sheet metal. Exposed nails shall be permissible only in top courses where unavoidable.

(d) Neatly fit slate around any pipes, ventilators, etc.

(e) Nails shall not be driven in so far as to produce a strain on the slate.

(f) Cover all exposed nail heads with elastic cement. Hip slates and ridge slates shall be laid in elastic cement spread thickly over unexposed surface of under course of slate.

(g) Build in and place all flashing pieces furnished by the sheet metal contractor and co-operate with him in doing the work of flashing.

(h) On completion all slate must be sound, whole and clean, and the roof shall be left in every respect tight and a neat example of workmanship.

FLAT SLATE ROOFS

Serviceable and permanent flat roofs are readily obtained by the use of roofing slate. The durable and wearing qualities of slate for promenade roofs are familiar to all who have seen and used slate for walks or floors. Slate possesses an apparent roughness due to slight irregularities of the "quarry cleft" surface and variation of color which give the appearance of a rough texture without actually being rough or hard to walk upon. As a result, where the design demands a promenade having qualities of this nature in addition to long life, slate will be found admirable. Its flexibility of arrangement adds to its other desirable features and allows unlimited possibilities in design.

Where the roof does not serve as a promenade, slate can be used also to advantage in place of the usual slag, gravel, slate chips or plastic slate covering of any built-up roof.

The nature of the material is such that it affords maximum protection against injury to the waterproofing membrane proper. The shape and weight of the units assure their being retained in place as first laid. The waterproof and non-absorbent characteristics of slate make it highly desirable as a surfacing material.

Slate

While any size slates may be used, the following have been recommended by the Division of Simplified Practice, United States Department of Commerce, as being readily available, economical and satisfactory:

Approximate thickness of slate pieces; in inches................ Commercial Standard, average or basic; 3/16″.
For promenade or extraordinary service: 1/4 and 3/8.

Dimensions of Slate Pieces; in Inches.		
6 x 6	10 x 6	12 x 6
6 x 8	10 x 7	12 x 7
6 x 9	10 x 8	12 x 8

For promenades the above or any size or shape required by the design may be employed.

While the commercial standard slates may be used for ordinary flat roofs, it is recommended that the 1/4″ thickness be used. For promenades it is advisable to use the 1/4″ or thicker. The 3/16″ slates afford ample wearing surface, but the 3/8″ will permit better bedding and remain more securely in place under traffic. Thicker slates may occasionally be required by unusual conditions or other than ordinary use. A leading manufacturer of built-up roofings recommends slates at least 3/4″ to 1″ thick, with sawed edges.

The slate should be ordered without punching or drilling. Slates less than 3/16″ thick are dressed or trimmed square with usual chamfered roofing slate rough edges. Slates 1/4″ and more may be either dressed or sawed edges if desired, although the latter is more expensive. When using Pennsylvania blue-grey slates for flat roofs clear stock should be specified.

Imbedding the Slate

There are several methods of imbedding the slate. Some roofers mop over the felt waterproofing with an elastic bedding compound of 60% pitch and 40% asphaltum.

In Newport, R. I., a manufacturer and roofer specializing in built-up roofs recently found it necessary to vary this mixture or top mopping coat to give satisfactory service to 50% Coal Tar Roofing Pitch and 50% Trinidad Roofing Asphalt. Another roofer has used Trinidad Steep Roofing Asphalt successfully.

A San Francisco roofer has found that all mastics made for other work would not answer for slate. The compound must cement the slate fast to the felt and yet not run in hot weather, nor crack in cold. It is poured on hot with a ladle and the slate pressed down into it firm so it fills joints solid. This concern then runs a cutter over joints to cut down surplus compound which extends above level and does not stick to surface of slates because they first apply a preparation thereon, so it is readily peeled off leaving the surface clean.

In several localities no special methods are employed. The three, four or five plies of felt are laid as recommended by their manufacturers, or in accordance with method of waterproofing employed, and standard thickness roofing slate are placed with butt joints in the top mopping coat instead of slag, gravel or slate chips, or other usual surfacing material. Many such roofs have been in service for several years. One case is known where standard thickness slate flat roof on a recreation pier has been walked on, danced on, and given very hard usage. For such installations of standard thickness slate, the metal legs of benches or chairs should be nailed to a wood strip in order to avoid damaging the joints or crushing through the slate edges and puncturing the waterproofing below.

Many successful installations have been obtained through use of cold plastic material for bedding slate. As a rule, about 10% of Portland Cement mortar or hydrated lime is added to the plastic to give it the proper stiffness and rigidity.

Another group recommending and using roofing slate for surfacing flat roofs claim only a plastic compound troweled on hot gives satisfactory bedding of slate if a plastic material is to be used.

A concrete setting bed is also recommended for promenades. Concrete composed of one part Portland Cement and three parts of sand or medium-size gravel is laid over the waterproofing felt. The concrete ¾″ in thickness forms a setting bed to receive the slate. After imbedding the slates and leveling them, grout the joints full or fill them with mineral wax. Expansion joints at intervals of 15 feet in each direction with suitable filler should be provided over the roof area.

Waterproofing

The supporting roof surface should be waterproofed by the method found by the designer to be most satisfactory in accordance with locality.

Where nailings may be necessary to hold felt in place, this should be done in such a way that the nails are covered with at least two layers of felt.

Metal flashings should be covered with one or two pieces of felt before the imbedding material is applied and the slate laid.

Roof Foundation

The supporting surface of the roof may be of wood or concrete. If of wood, the usual sheathing boards as described under sloping roofs may be used. For mill construction, use plank sheathing of proper thickness for the span or rafter spacing. Whether wood, concrete or other material is used for this purpose, it is important that the surface be fairly smooth and free from nails or other projections which might puncture the felt or wear through under traffic. The sheathing slab or filling under the waterproofing should be sloped to the roof drainage points so that any possible water getting under the waterproofing from seepage of coping wall joints, etc., may be readily carried off.

Roof Slope.

The slope for roofs of this type should not exceed 4″ rise in 12″ of run. Slopes of less than this amount are recommended. For most conditions, ½″ or ¾″ per foot will be found ample and as little as ¼″ will be satisfactory. When the slope is less than ½″ per foot, it is important to see that the surface is evenly maintained and hollow spots avoided.

Weight of Roofing

The weight of the roofing material above the roof slab or boards is given in the following tables:

WEIGHT OF FLAT SLATE ROOF WITHOUT CONCRETE BEDDING SLAB

Materials	Weight of Materials per sq. (100 sq. ft.)	Total Weight per sq. (100 sq. ft.)	Total Weight pounds per sq. ft.
Waterproofing (Weight varies, assumed here to be 150 lbs.)	150	....	....
3/16″ Slate	250	400	4.0
¼″ Slate	335	485	4.85
⅜″ Slate	500	650	6.5
½″ Slate	675	825	8.25
¾″ Slate	1,000	1,150	11.5
1″ Slate	1,330	1,480	14.8

WEIGHT OF FLAT SLATE ROOF WITH CONCRETE BEDDING SLAB

Materials	Weight of Materials per sq. (100 sq. ft.)	Total Weight per sq. (100 sq. ft.)	Total Weight pounds per sq. ft.
Waterproofing (Weight varies, assumed here to be 150 lbs.)	150	....	....
¾″ Concrete Bed	750	....	....
3/16″ Slate	250	1,150	11.50
¼″ Slate	335	1,235	12.35
⅜″ Slate	500	1,400	14.00
½″ Slate	675	1,575	15.75
¾″ Slate	1,000	1,900	19.00
1″ Slate	1,330	2,230	22.30

FLAT SLATE ROOFS—SPECIFICATIONAL DATA

There are many methods for laying built-up roofs. There are many more opinions as to the most satisfactory manner of imbedding the slate or tile on top of the built-up roof whether or not for promenade purposes. It has therefore seemed best in this edition to present the experience and specification data of some successful flat roof projects where roofing slate has been used as the surfacing material. Information on the use of slate chips instead of gravel or furnace slag or plastic slate surfacing will be furnished on request by the Association. As this volume deals only with roofing slate units, such types of slate roofs are not discussed. Nor are those roofs where slate granules or laminated slate surfaced roll roofing or individual shingles described herein. A list of members furnishing crushed slate for such purposes can be furnished on request.

Every architect, builder or owner and even roofers in different localities have found by experience one or another method of built-up roof construction more suitable.

Possibly as an outgrowth of the Federal Specification Board for unified, simplified, practical government specifications, and through closer co-operation of roofing contractors through associate membership in this organization, there may appear in later editions of this work a standard specification for placing roofing slate on flat roof surfaces. Any architect, builder or owner desiring the protection advantages of roofing slate for the waterproofing membrane of flat roofs can obtain a list of similar projects in their vicinity so as to inspect them if they so desire before writing their specifications. Flashings and complete drainage of all water falling on the roof area or from seepage around stone work, etc., at coping or other edges of waterproofing base or the finished roof surface are prime requisites for satisfactory service from any flat roof construction. As one expert well states it:

> "Flashings and counter flashings seem to be the source of the greatest trouble in connection with roofs of all kinds. A great many people who design the construction upon which roofs must be placed seem to give no consideration to the limitations to the different materials that are used. They do not realize that the changes in temperature and the frosts very materially effect flashings, etc., and I have often remarked that it is rather strange that so little progress has been made up to the present time in connection with roofs, which are probably one of the first things that man ever commenced to construct in connection with habitations."

Some details of proper flashing for slate flat roofs are shown on page 23 as adapted from Copper and Brass Research Association handbook on Flashings. With the development of real research by such associations and the National Slate Association, not only will progress be made regarding roofs but many of the avoidable troubles will be done away with if recommendations are followed.

The following are typical specifications for flat slate roofs which may be helpful or suggestive for wording others. The National Slate Association particularly requests experience data under such specifications or any others which have secured satisfactory flat slate roofs, whether or not for promenade purposes, so that eventually its standard Specification "F" may be evolved for such roofs by the Slate Industry.

A roofing concern in New York City which has specialized with great success on flat slate roofs publishes this specification:

Specifications for Flat Slate Roofs over Concrete or Board Construction

> "After the roof surfaces have been properly graded to the proper outlets by other contractors, coat this construction with a heavy coat of Pure Asphalt Bitumen Roofing Cement, applied hot, in which lay shingle fashion five (5) layers of Standard Roofing Felt sticking the entire width of the sheets with the above asphalt.
>
> "After sixteen (16) ounce copper or other metal base flashings have been set by Metal Contractor extending four inches (4") on roof and nailed three inches (3") on centers, connect these flashings to the five (5) ply waterproofing with one (1) strip of felt and hot asphalt. As a walking or wearing surface imbed in a heavy flowing coat of Pure Asphalt Bitumen Steep Roofing Cement 6" x 6" Best Quality Slate not less than 1/4" thick.
>
> "P. S.—If this roofing is applied over boards, omit the first coating of asphalt and blind nail the back edges of felt."

The reflecting pool basin of the Lincoln Memorial is waterproofed and lined with roofing slate. Regular built-up construction is used, and after bedding the slates in the top mopping coat, they were again mopped on top so that there really is a coating both above and below the slate.

Excerpts from Specifications for New Roof on Treasury Building

> "The roof covering not otherwise specified or indicated shall consist of the following minimum plies or layers and amounts of materials:

1 layer of sheathing paper per 100 square feet, 5 pounds.
5 layers of roofing felt per 100 square feet, 75 pounds.
4 layers of roofing cement per 100 square feet, 120 pounds.
1 layer of bedding compound per 100 square feet, 75 pounds.
1 layer of slate.

"Where the roof covering is applied directly over concrete the sheathing paper shall be omitted and 5 layers of roofing cement will be required with a weight per 100 square feet of 150 pounds.

"Bedding compound shall be thoroughly plastic when heated for application, and must be strongly adhesive, free from any tendency to brittleness or cracking, or stickiness or creeping under natural temperatures when in place, and be unaffected by the elements. The compound shall not run or creep on a slope of 4½″ to the foot at a temperature of 140° F. when tested on glass or a similar smooth surface, and shall not lose its elasticity at a temperature of 0° F.

"Slate shall be sound, hard, of fairly uniform thickness, not less than 3/16″ thick; with square-cut edges, and in sizes generally about 8″ x 10″. Smaller sizes shall be used on curved surfaces as necessary.

"The slate shall be solidly bedded in hot bedding compound with butted joints, and neatly fitted to all connecting work. The slate shall be clean and dry when laid. The completed roof shall be smooth and regular, and free from cracked, broken or loose slate."

One roofing contractor associate member specializing on surfacing flat roofs writes:

"We have applied several thousands of squares of flat slate roofs in the past twenty-five years. The usual procedure is first to lay a three, four or five ply tarred felt and pitch roof, laid according to the manufacturer's specifications, and over this roofing imbed, in a mixture of 60% coal tar pitch and 40% Trinidad asphalt, Standard Stock roofing slate.

"After several years of experience we are firmly convinced that the most serviceable specification for a promenade deck should be the Standard Specification of any of the well-known built-up roofing manufacturers, surfaced with sawed-edge slate slabs approximately ⅞″ thick, laid in cement mortar over the roofing specifications above referred to."

A concern specializing in the construction of mausoleums has used various materials for roofing flat surfaces. Finding that maintenance and repair expenses varied for all types of flat roofs a special investigation and research, costing from three to four thousand dollars, was undertaken. After investigating all roofs of their own construction as well as many flat roofs built by others it was found that one with slate surface was still intact and had required no upkeep or repairs after 10 years, even though it was found that many large stones had been thrown on the slate. Slate was adopted as standard for flat roof surfacing and the method used by the concern is outlined as follows:

"The Slate Roof that was built some ten years ago, was built with the standard built-up roof construction.

"The slate were imbedded in the top mopping coat instead of slag. We have used no special method. Our latest building has been built in the same manner, and since completed has been examined by a number of roof builders who pronounced it as a roof that excels any other roof construction."

There is another group, developing built-up roofs and asphalts for flat slate roofs, who have this to say:

"The admixture of asbestos fibre to a soft asphalt is better practice than to harden the asphalt itself in the process of refining. A soft asphalt reinforced with asbestos fibre is still a soft, adhesive asphalt and retains its adhesive qualities despite the fact that it has been stabilized and reinforced with the fibre. For the particular purpose of serving as a bed for slate it can be still further improved by the admixture also of slate flour. The asbestos fibre reinforces it but the slate flour gives it body and makes it easier to handle on the job, the same way that mortar is easier to spread than a liquid pitch."

From the foregoing it will be seen that it is impossible at the present time to include here a recommended specification, or even suggested paragraphs, for a flat slate roof. The Association is collecting data continuously on this subject, and will be glad to furnish the latest available information to architects and contractors upon request, and experience will be welcomed.

TYPICAL USES OF ROOFING SLATE FOR FLAT SURFACES

Usual Flat Slate Roof

Standard Thickness Slate Promenade, Recreation Pier

Mausoleum with Flat Slate Roof

THE LINCOLN MEMORIAL
Roofed with Square Thick Roofing Slate with Sawed Edges

REFLECTING POOL
Lined with Regular Roofing Slate

Showing the Roofing Slate Lining

Roof of Mausoleum Shown Above

Reroofing with Slate

IN old Oriental rugs, rare oil painting or ageing houses alike, we are consciously or unconsciously aware of the effect of time. The contrasting colors of the rug are softened, those in the painting slightly dimmed, and the house has been toned so that the materials blend as one. We would instantly resent the insertion of fresh, new colors in the rug or painting. Yet some people thoughtlessly reroof their houses with a roofing material which cannot help but be incongruous and unharmonious with its surroundings. The ageing exterior walls and the profuse foliage of the trees and vines call for the use of materials which quickly become a part of the composition.

Slate possesses qualities which make it at once suitable for reroofing any type of building. It is a product of nature, obtainable in neutral or colorful, permanent or weathering tones, which quickly assumes the characteristics of age, blends with the natural surroundings, or by proper selection will conform to any desired color scheme. Its texture meets the demands of any design. Its cost is only slightly more than the less permanent roofings.

Slate, being a hard, dense stone, is fireproof, stormproof, permanent, and cannot warp or curl. Reroofing with slate settles the replacement problem for the last time and assures economy in maintenance and lowest insurance rates.

A reasonably smooth under surface is essential to the proper laying of any roof covering. Where the old roof covering presents a fairly even surface or can readily be made so, the slate may be laid directly on top of it. This procedure has been common practice for years, produces satisfactory results, and the old roofing possesses some insulating value.

For high-grade work, it is strongly recommended that the old roof covering be removed. This insures proper examination of roof construction, rafters, etc. It permits of moving shingle laths or roof boards to proper spacing for slate, renewing broken ones and nailing all solid. In sections where heat conservation and the keeping out of air is an advantage, it allows the filling in with new roof boards between the lath so that the under surface may be solidly sheathed, smoothed and covered with at least a 14-pound asphalt felt. Should a respon-

Applying Slate Over the Old Roof

sible slate roofing contractor's inspector recommend the removal of the old covering, the cost of this adds only a small amount to the entire cost of the new slate roof.

For a slate roof of commercial standard thickness slate, roof construction conforming to good engineering practice for roofs of lighter materials will usually be found adequately strong. This applies both for new construction and for reroofing, and will be found discussed in detail on page 34.

The reroofing data which follows was originally published by the National Slate Association under the title of "Hints to Roofers on Reroofing with Commercial Standard Slate." Since much of the information therein must be considered by the architect in recommending a new roof covering in alteration work and by the owner in deciding on the type of roof, it is reprinted here in the hope it will prove of value.

When the Old Covering Is Removed

The removal of the old covering has the advantage of exposing the sheathing or roof lath and permitting a thorough inspection. Any broken boards or lath should be replaced with new whole material and all loose boards should be securely nailed in place. Boards originally so laid that the joints were not broken over bearings should be taken out and replaced or short rafters or blocking cut in to act as a bearing. In localities where the saving of heat is essential and roof lath has been used, if the lathing is not to be removed, new boards should be used to fill the space between the old lath. Both should be of the same thickness to provide a reasonably smooth surface.

It is important to go over the roof and remove or drive "home" any projecting nails and cut down any warped or raised edges or ends of sheathing or lath. Before laying the felt, thoroughly sweep off the sheathing to remove all chips, blocks and loose nails.

When the Old Covering Remains

Where the old roof covering is allowed to remain, see that any low spots or loosened areas are filled up and made secure. Where possible to do so, inspect the sheathing or lath from the under side and repair all broken or loosened boards. In ordering nails, make allowances for the thickness of the old roofing and use a nail of sufficient length to secure thorough penetration into the roof boarding.

Note the width of the roof boards or lath spacing and order slate of proper length to secure a nailing which will avoid the joints. The following table will be useful in this connection:

Spacing of Lath (c. to c.)	Length of Slate
10½ inches	24 inches
9½ inches	22 inches
8½ inches	20 inches
7½ inches	18 inches

Heavier Rafters Are Not Required

Rafters adequate for wooden shingle roof will be of sufficient strength for slate of commercial standard (3/16") thickness.

Laying Slate Over Old Roofing Material

It will sometimes be found that the rafters of old buildings—and some not classed as old—while of ample strength to support the present roof covering, were not designed to carry the additional snow or wind load recommended by present-day engineering practice or required by the local building code.

When existing roof appears to sag or gives indication that existing supports are not adequate for the present roof covering, it would be unwise to replace the old covering or cover with any new material without strengthening the roof supports.

To aid architects and owners in overcoming such conditions, we would be glad to advise methods or send blue prints of drawings which the Structural Service Bureau has prepared showing methods for strengthening roof construction. It rarely occurs that conditions are such that the entire roof construction must be removed and replaced with members of larger size or different arrangement.

How a Northern Roofer Lays Slate Over Shingles

"In regard to the over-shingles work, will say that it works out very satisfactorily. Occasionally I have a job which is badly cut up with hips, dormer windows, etc., which should be stripped, and I do that. But the average roof lays very nicely over shingles with slates 18″ or longer, and valleys go in well over shingles. Short slates do not lay well over shingles, as the slate necessarily rests on the shingle butts and it should span at least two courses of the wood shingles.

"I lay the American method entirely, and do over-shingles work for about 25% more than the cost of stripping and reroofing with wood shingles. A 6d nail is used, which goes through the old shingles and into the roof boards. I also punch four holes in the slate instead of only two, as sometimes we don't get good nailing and have to use the other pair of holes."

Another roofer, who has made a specialty of over-shingle work secures better results where the old surface is more or less uneven, with slates 12″ and 14″ long. He says, "They lay well and fit the contours much better than the larger sizes and further they are less liable to breakage where carpenters, painters and others must use the roof." The Association partial payment plan makes reroofing with slate conveniently financed.

MAKING ADDITIONS OR ALTERATIONS TO SLATE ROOFS

Once a slate roof is properly laid it will be permanent and require little or no upkeep or care. However, houses are sometimes enlarged or remodeled. In such cases it is often necessary to join a new and old roof or to remove and alter sections of the existing roof.

It is desirable and necessary that the altered or additional roof match the existing one in both shade and texture. To obtain this result, it is advisable to secure slate of the same quality and color as the original slate. Slates from some quarries weather, that is, the color as first quarried will differ from the permanent shade resulting from a few months' exposure to the weather. Other slates are permanent or unfading and do not mellow on exposure but retain the original natural shades.

To match slates that are unfading, requires unfading slate of the same shade or slate which will weather to the desired shades. If the slate on the roof has already weathered, it can be matched with unfading slate or with weathering slates which will mellow to the desired effect. Securing slate from the original quarry, reduces trouble about matching or colors. However, in many remodeling jobs, the slate has been on the roof for many years and no record of the quarry from which it came will be available.

An experienced slate roofer can usually identify slate as to producing vein but in case of doubt samples may be submitted to the Association for classification.

The best method of procedure is to remove small adjoining sections and relay, mixing some new slate with the old. This will prevent a clear line of demarcation where the new work adjoins the old and the completed roof will at once present a satisfactory appearance.

In minor alterations such as adding or removing a dormer, the old slate which is removed can be used again. In dormers and other projections, the lights and shadows will differ from those on an expanse of roof, making it easier to add new slate which will be unnoticed. For example, new slates could best be used on cheeks of a new dormer using old slates on other parts. Some roofers buy up a number of old roofs from buildings which are being torn down and thus obtain old weathered slate in their yards which can be readily matched with the slate on the roof when minor alterations are made.

However, due to unavoidable causes, slates are sometimes broken on the roof. The broken slate can be repaired by the Slate Roofer. The best method is to first remove the broken slate, cut the nails with a ripper and remove any remaining small pieces of slate. Insert new slate and nail this slate through the vertical joint of the slates in the overlying course approximately 5 inches from the head of the slate, or 2 inches below the tail of the second course of the slate above; over this nail insert a piece of copper approximately 3 inches in width by 8 inches in length. The piece of copper should be inserted under the course above, lengthwise, so that it will extend a couple of inches under the succeeding course, thus insuring a proper lap and protection throughout the exposed joint in which the nail is driven. This small piece of metal should be first bent slightly concave or convex which will insure its remaining tightly in place.

When making alterations to a slate roof, only responsible and experienced slate roofers should do the work. If other workmen are required to use ladders or scaffolds on slate roofs, boards should be used under the legs or uprights to distribute the pressure.

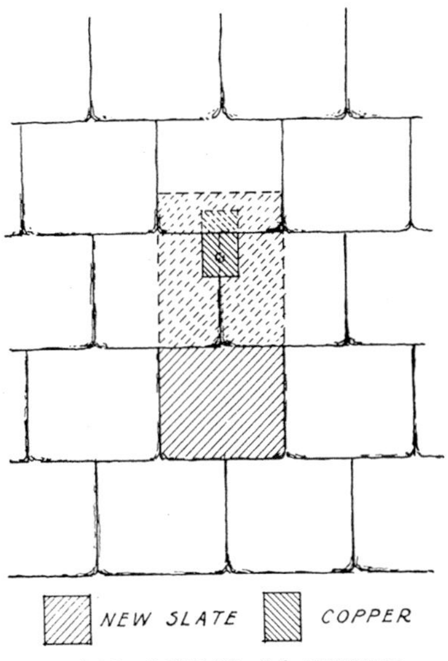

PROPER METHOD OF INSERTING A NEW SLATE

MISCELLANEOUS METHODS OF LAYING SLATE

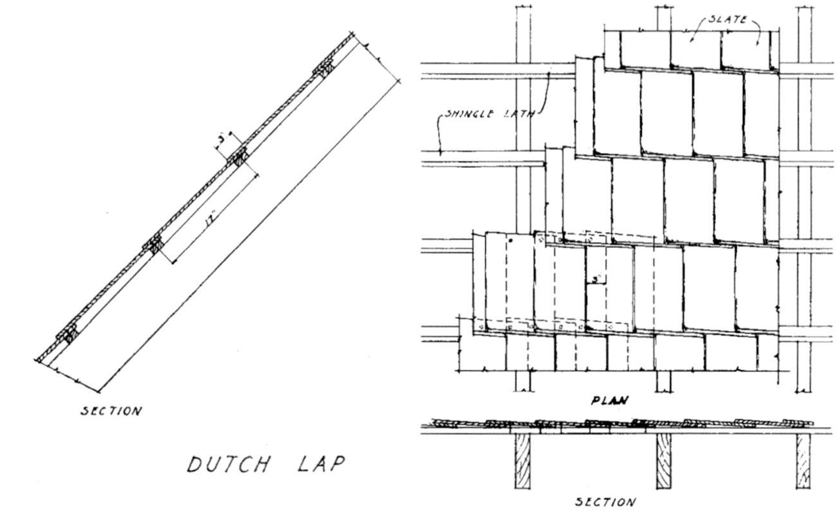

DUTCH LAP

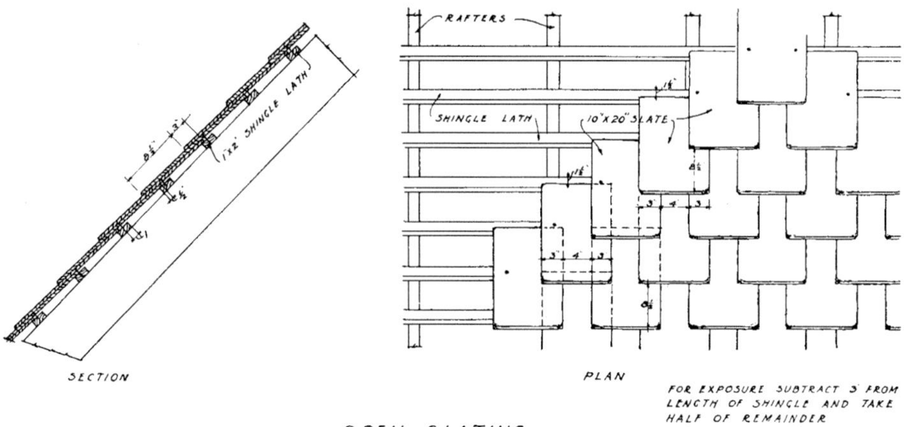

OPEN SLATING

FOR USE WHERE VENTILATION IS DESIRABLE

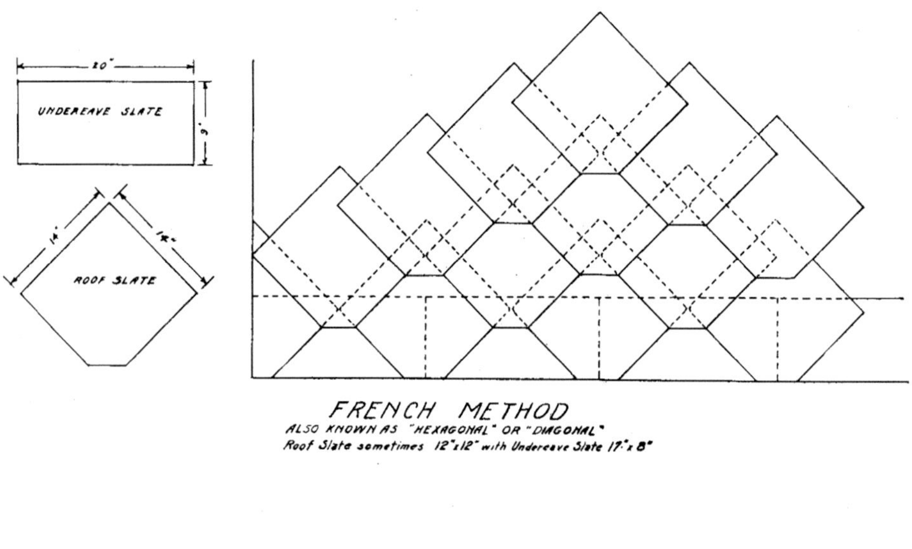

FRENCH METHOD

ALSO KNOWN AS "HEXAGONAL" OR "DIAGONAL"
Roof Slate sometimes 12"x12" with Undereave Slate 17"x 8"

Three methods of laying slate which are less expensive than the standard types, are here illustrated as they are in more or less common use in certain localities.

When properly laid they seem to prove satisfactory for certain purposes and locations.

Dutch Lap:
Laid with regular slate on shingle lath or tight sheathing.

Open Slating:
This method is also especially suitable for barns and other buildings or where ventilation is desirable such as slaughter houses or in mild climates.

French Method:
The diagonal, French or Hexagon method of slating being less expensive than the standard American method, has appealed mostly to New England thrift. Many, many carloads of Maine and Vermont roofing slates have been so laid in certain New England cities not only when reroofing with slate over old temporary roof coverings, but also on new roofs. When laying new roofs of this method in that section where heat conservation is a factor, thirty pound instead of fourteen pound asphalt saturated felt is first applied on solid roof deck of sheathing or roof boards. The two sizes shown in the last figure on this page, with their respective undereave slates, are usually furnished and stocked only by certain quarries. For greatest service as with standard slate roof, sufficient lap of approximately 3 inches must be provided. The 14 inch by 14 inch size is used on houses or buildings with large gables and sections, while the 12 inch by 12 inch are used on smaller ones.

GENERAL DATA ON SLATE

THE history of slate is an excellent criterion of the serviceability of this material for roofing. In France, England and Wales can be found better evidences of the durability of slate for this purpose than in our own country, for there slate roofs have been used for centuries.

One of the earliest references to this use of slate is the construction of the roof of the Saxon Chapel at Stratford-on-Avon, Wiltshire, England. This was built during the eighth century and after 1100 years of constant exposure to all kinds of climatic changes, it is still in good condition although covered with moss.

Slates were used to cover old castles at Carnarvon and Conway in North Wales during the twelfth century.

Typical Slate Roofed Railroad Station

These slates were thick and rough, as the workers had little skill in splitting and trimming. Today modern architecture is asking craftsmen to reproduce the handiwork of these early workers on special treatment, rough and thick slate roofs.

The first authentic reference in literature to slate quarrying is the mention of the Penrhyn quarries by a Welsh bard in 1570. In 1580 Sion Tudor addressed some verses to the Dean of Bangor, Rowland Jones, requesting a shipload of slate. It is fitting that the introduction of the world to so vast an industry should have come through the agency of people of such importance as the Dean of the Church and a poet of distinction.

During the next two hundred years little is known of the growth of the industry, but gradually from a local market about Penrhyn, it had grown to considerable importance by the end of the eighteenth century. Quarry methods were crude and transportation difficult until 1850 when the growth of the Welsh slate industry became very rapid due to the extensions of railways and the resulting widening of the market, the growth of new towns, improvements in rural dwellings, removal of restrictive tariff and increased foreign demand.

In France the rapid development of the slate industry dates from about this same year (1850) and for nearly the same reasons. The quarrying of slate on the continent had begun long before this, however, and at Angers, France, a famous modern slate mining center, a slate roofed castle built about the twelfth century is still standing.

A recent use of slate in Europe which is of interest to all Americans is the roof of the library of the University of Louvain. The Library, which was destroyed by the Germans in 1914, is being reconstructed by funds contributed by the Schools, Colleges and Universities of the United States. The plans were drawn by American architects and the work of construction is being done by an American firm. The roof will be of blue slate providing an excellent background for the elaborate carving which will embellish the central motive and balustrade of the facade.

Early History in America

In America probably the oldest slate quarry is in the Peach Bottom district at the Pennsylvania-Maryland line.

Suggesting a Picturesque Use of Slate Roof and Trim, Fifth Avenue, New York

Regarding these quarries, Ferguson, writing in 1910, says that slates mined near Delta in 1734 were at the time of writing, 1910, still in use covering the seventh building on which they had been placed, showing no indication of change in color or deterioration in quality.

In Virginia, the first quarry was opened about 1787 to provide slate for the roof of the state capitol. Slate was first quarried in Georgia in 1850. From such beginnings, slate quarrying grew and spread until it became an established industry. From the very first, Welsh slate workers had an important part in the development of slate quarrying in America. It is a matter of record that about 1877, one hundred and fifty skilled slate workers left Bethseda district in North Wales to work in American quarries. Many years prior to that, however, at the very beginning of the industry, Welsh slate workers were employed as is indicated by the names of many places in the slate regions, as Pen Argyl and Bangor, in Pennsylvania, which were named after Welsh towns.

To attempt to list here all the prominent structures in America which are protected by roofs of slate would be obviously impossible. The National Slate Association maintains a list of such buildings and will be glad to refer architects to work in their locality where the various types of slate roofs and other features may be examined.

Foreign Slate Deposits

In the British Isles, the Welsh slates, particularly, those of Carnarvonshire and Merionethshire, are the best known, though slates are also produced from Cornwall, England, from Argyle, Perth, and Dunbarton, Scotland, and from Tipperary and Cork Counties, Ireland.

Stable and Gate Lodge Roofs and Pigeon Runway All of Roofing Slate

Attractive Community Theatre, with Roof, Sidewalk and Lobby Floors of Slate

The chief slate mines in France are in the Ardennes and the Angers Districts. Great quantities of slates have been taken from these regions and much of the product has been exported. Slates have been quarried in Bohemia and near Olmutz in what was formerly the Austrian Empire, and along the Valley of the Moselle and in Westphalia, Germany.

A limited production of slate has been noted from Norway, Portugal, India, Italy, New Zealand and Tasmania.

Some of these foreign slates like some American slates formerly quarried are not true slate in important mineral constituents. The Association will gladly report on any specimen sent it.

Geological and Chemical Formation

The following definition of slate has been tentatively adopted by Committee D-16 on Slate of the American Society for Testing Materials.

"Slate is a microgranular crystalline stone derived from argillaceous sediments by regional metamorphism, and is characterized by a perfect cleavage quite independent of original bedding, which has been induced by pressure within the earth. The essential mineral constituents of slate are white mica (chiefly sericite) and quartz. Secondary prominent constituents are black mica (biotite), chlorite and hematite. Minor accessory minerals are carbonates, magnetite, apatite, clay, andalusite, barite, rutile, pyrite, graphite, feldspar, zircon, tourmaline and carbonaceous matter. Igneous slates, because of their rare occurrence and insignificant commercial importance, are excluded from this classification.

"Slate for roofing purposes should contain not more than x percent of carbonates (calcium or calcium-magnesium carbonates?).

"Slate for electrical purposes should contain not more than x per cent of metallic oxides or carbon."

In the last two paragraphs the percentages are still

QUANTITATIVE CHEMICAL ANALYSIS

ORIGIN	COLOR	QUARRY	SiO_2	Al_2O_3	Fe_2O_3	FeO	CaO	MgO	K_2O	Na_2O	CO_2	FeS_2	C	H_2O	TiO_2	S	MnO	BaO	Authority	SPECIFIC GRAVITY
MAINE	BLACK	MONSON	56.42	24.14			0.52												L.M.NORTON	2.85
MICHIGAN	DARK GRAY	L'ANSE	57.90 58.40 58.10	18.84 18.52 19.40	9 9 9	90 10 35	1.26 1.46 1.18	3.80 4.10 4.05	3.75 3.75 390	0.30 0.32 0 36	TRACE TRACE TRACE	TRACE TRACE TRACE	TRACE TRACE TRACE	0.25 0.42 0.51	0.66 0.82 0.74				H.V.ANDERSON R.NOTVEST R.NOTVEST	3.1
NEW YORK	RED	GRANVILLE	67.55	12.59	5.61	1.24	0.26	3.27	4.13	0 61	0.11	0 04		303	0.58		0 19	0.31	G.STEIGER	2.789
PA.	GRAY	PEACH BOTTOM	58.37	21.99	10	66	0.30	1.20	1	93	0.39		0.93	4.03		0 10			GARRET BOOTH BLAIR	2.894
PA.	GRAY	PEN-ARGYT	57.50	20.84	1.30	2.21	3.92	3.02	3.36	1.25	3 30	1 39	0.57	0.15					H.V.ANDERSON	2.813
PA.	GRAY	SLATINGTON	56.10	15.15	1.67	2.90	4.05	2.82	3.50	1.30	3.60	1 72	0.59	4 10	0.78		0 9		R.NOTVEST	2 78
PA.	GRAY	SLATINGTON	56.85	15.24	5	52	4.24	2 93	3.34	1 38	3.58	1.72			0.84		0.9	0.8	W.F.Hillebrand	2 78
CALIFORNIA	GREEN	EUREKA	47.30	15.53	8	00	7 83	7.86	3	17	9.92					0 12			W.T.SCHALLER	
GEORGIA	GRAY GREEN	BOLIVAR	56.02	21.61	1.36	5.97	1.22	2 96	2 83	1 26	0.91				0 74	0 35	0 05		E.EVERHART	
NEW YORK	GREEN	GRANVILLE	67.89	11.03	1.47	3.81	1.43	4.57	2 82	0.77	1.89	0.04		3.21	0.49		0.16	0.04	W.F.Hillebrand	2.717
VERMONT	SEA GREEN	WEST PAWLET	67.76	14.12	0.81	4.71	0.63	2 38	3.52	1.39	0 40	0.22		2 98	0.71		0.10	0.04	W.F.Hillebrand	2 791
VERMONT	UNFADING GREEN	POULTNEY	59.48	18.22	1.24	6.81	0.56	2 50	3 81	1.55	0.39	0 13		4 05	1 02		0 07	0.05	W.F.Hillebrand	2 795
VERMONT	PURPLE	HYDEVILLE	60.96	16.15	5.16	2 54	0.7	3 06	5.01	1 50	0.68			3.08	0 86		0.07	0.04	W.F.Hillebrand	
VERMONT	MOTTLED PURPLE GREEN	POULTNEY	60.24	18.46	2.56	5.18	0 33	2 33	4.09	1.57	0.08	0 16		3 81	0 92		0 07	0.03	W.F.Hillebrand	2 805
VIRGINIA	DARK BLUE	ESMONT	56.33	22.26	9	44	0 68	1 48	5	07	1.69			2 86		0.04			LEHIGH TESTING	2 90
VIRGINIA	DARK GREEN	ESMONT	53.58	24.53	8	81	0.42	1 95	4	88	1.99			3.39		0.08			LABRATORIES ALLENTOWN	2 90
PA.	RIBBON	BANGOR	55.10	18.95	1.45	2.30	2.88	3.06	3.32	1.28	3.15	1.62	4 10		0.88				WIRTH & NOTVEST	2.953
PA.	GRAY	BANGOR	57.10	21.05	1.35	2.05	3.86	3.10	3.27	1.31	3.30	1.35	0.6		0.45				WIRTH & NOTVEST	2 872
PA.	GRAY	WINDGAP	57.30	21.05	1 30	2.28	3 85	3 15	3 32	1.28	3 40	1.42	0 53		0.36				WIRTH & NOTVEST	2 835

FOR DETAILED PROCEDURE OF ANALYSIS SEE U.S. GEOLOGICAL SURVEY BULLETIN # 700, "THE ANALYSIS OF SILICATE & CARBONATE ROCKS" by DOCTOR W.F. HILLEBRAND the subject of consideration. The National Slate Association will be glad to advise regarding the suitability of the mineral constituents of any slate for a particular location.

Origin of Slate

These minerals constituting slate were deposited in bodies of water and erosion and deposition accounts for the different composition in successive beds. Other materials may have been deposited over the clays, and the pressure of the superimposed material may have gradually united the clays in shale. Shale has been described as "a laminated rock consisting essentially of clay, but it does not possess the splitting properties of slate." Some of the beds of shale remain as such today, but many have been subjected to the intense pressure and high temperature of the crumpling and folding of the earth's crust. When this has been the case the shales have been transformed into slates.

A Slate Quarry

This tremendous pressure changes the position of the mineral grains until they lie parallel and at a definite angle to the direction of pressure. The high temperature is at the same time acting with the pressure tending to change the constituent minerals to new minerals, such as mica, quartz, chlorite, magnetite, graphite and others. The "slaty cleavage" or ease of splitting in one direction is the result of the parallel position of the mineral grains.

The Mineral Constituents of Slate

The table on page 75 shows the mineral constituents of slate and their important physical characteristics. A petrographic analysis of slates or any stone or rock formation is the only true study to determine the mineral constituents which are the real proof of the identity of any stone or its classification among natural rock formations.

The table on page 73 shows the results of a quantitative chemical analysis of slates from various parts of the country. Chemical analysis figures are valuable for comparative purposes, but often misleading because the same elements may be contained in varying proportions in minerals of very different chemical and physical characteristics.

The analysis of Pennsylvania hard-vein slate is given separately because it has been made since the table on page 73 was prepared.

Moisture at 100° Centigrade		.048%
Silica	(SIO_2)	61.27%
Ferric Oxide	(Fe_2O_3)	5.94%
Aluminum Oxide, etc	(AL_2O_3)	16.46%
Manganous Oxide	(MnO)	.044%
Calcium Oxide	(CaO)	2.76%
Magnesium Oxide	(MgO)	3.399%
Potassium Oxide	(K_2O)	3.73%
Sodium Oxide	(Na_2O)	5.92%
Total Sulphur as Sulphur Trioxide	(SO_3)	.84%
Total		100.411%

The Quarrying of Slate

Slate is quarried by different methods depending upon the structure and bedding of the rock.

In Pennsylvania, the deposits are in nearly vertical beds, but curve back and forth at steep angles. In the New York and Vermont slate area near Granville, the bedding dips at an angle ranging from 15° to 60°, the average being between 40° and 45°. Here, therefore, the plan of quarrying is different from that in Pennsylvania.

The vertical beds of Pennsylvania permit a small surface opening which can be worked to great depth and operated for many years without the expense of removing the overburden. Where the beds are at an angle it is necessary to "strip" or remove large quantities of waste rock or earth.

Separating Block from Floor

The OPTICAL & PHYSICAL CONSTANTS of the MINERAL CONSTITUENTS of the SLATES.

NAME	CHEMICAL COMPOSITION	OPAQUE COLOR BY INCIDENT LIGHT	ISOTROPIC	ANISOTROPIC: COLORLESS UNI	ANISOTROPIC: COLORLESS BIOXIAL	ANISOTROPIC: COLORED NON-PLEOCHROIC UNI	ANISOTROPIC: COLORED NON-PLEOCHROIC BIOXIAL	ANISOTROPIC: COLORED NON-PLEOCHROIC UNI	ANISOTROPIC: COLORED NON-PLEOCHROIC BIOXIAL	FORM OF OCCURENCE	REFRACTIVE INDEX. AVERAGE	MAXIMUM BIREFRINGENCE	PLEOCHROISM	ORIENTATION	OPTIC AXIAL ANGLE	HARDNESS MHOS SCALE	SPECIFIC GRAVITY	MICRO-CHEMICAL REACTIONS
PYRITE	FeS_2	BRASS YELLOW	—	—	—	—	—	—	—	CRYSTALS & IRREGULAR GRAINS	—	—	—	—	—	6-6.5	4.9-5.2	NON-MAGNETIC, INSOLUBLE IN HCL. READILY SOLUBLE IN HNO_3, WHEN TESTED GIVES OFF SO_2 FUMES.
PYRRHOTITE	FeS	BRONZE YELLOW	—	—	—	—	—	—	—	GRANULES	—	—	—	—	—	3.5-4.5	4.4-4.6	MAGNETIC
MAGNETITE	Fe_3O_4	BLACK	—	—	—	—	—	—	—	OCTAHEDRONS, CUBES IRREGULAR GRAINS	—	—	—	—	—	5.5-6.5	4.9-5.2	STRONG MAGNETIC, NOT READILY SOLUBLE IN HCL
GRAPHITE	C	BLACK	—	—	—	—	—	—	—	FLAKES-NEEDLES	—	—	—	—	—	3.5-4.5	4.4-4.6	INSOLUBLE IN ACIDS, CONTAINS OFTEN Fe_2O_3 AND Al_2O_3
HEMATITE	Fe_2O_3	RED RARELY CRYSTALIZED						RARELY HEXAGONAL NEG.		HEXOGONAL CRYSTALS OR AS RIM AROUND MAGNETITE	IF CRYSTALIZED 2.50	0.250	O = BROWNISH RED E = YELLOWISH RED			5-6.5	4.2-5.3	NOT READILY SOLUBLE IN ACIDS, LIMONITE IS YELLOW, MAGNETITE BLACK, HEMATITE RED.
APATITE	$3(Ca_3P_2O_8)Ca(Cl\,F)_2$	COLORLESS	COLORS IN THIN SECTIONS	HEXAGONAL NEG.						SMALL PRISMS LARGE CRYSTALS	1.63	0.004	SOMETIMES PLEOCHROIC			5	3.2	READILY SOLUBLE IN H_2SO_4, THIS SOLUTION GIVES YELLOW PRECIPITATE WITH AMMONIUM MOLYBDATE.
QUARTZ	SiO_2	COLORLESS		HEXAGONAL POS.						GRANULES CRYSTALS	1.54-1.58	0.009	SOMETIMES PLEOCHROIC		ANOMALOUS 2E = UP TO 25°	7	2.5-2.8	INSOLUBLE IN COMMON ACIDS, SOLUBLE ONLY IN HF, SiO_2 AS Na_2SiF_6
ZIRCON	$ZrO_2\,SiO_2$	COLORLESS		TETRAGONAL POS.	↓					SMALL STOUT CRYSTALS	1.83-1.99	0.050 to 0.062	VERY WEAK BUT BRILLIANT INTERFERENCE COLORS.			7.5	4.4-4.7	NOT SOLUBLE IN ACIDS, ONLY HF.
KAOLIN	$H_4Al_2Si_2O_9$	COLORLESS AMORPH WHITISH			IF CRYST. NEG. MONOCLINIC					FLOUR-LIKE WHITE OPAQUE FLAKES SELDOM IN SCALES	1.56-1.57	0.005		IF CRYSTALIZED EXTINCTION ANGLE = 13°		1-1.5	2.2-2.6	NOT READILY SOLUBLE IN ACIDS, HIGHLY REFRACTORY, GIVES OFF WATER ABOVE 330° C.
FELDSPAR	ANORTHITE - PLAGIOCLASE $CaO\text{-}Al_2O_3\text{-}2SiO_2$	COLORLESS			TRICLINIC NEG.					CRYSTALS GRANULES	1.53-1.54	0.005			2V = 43°-53°	6-6.5	2.7-2.8	SOLUBLE IN HCl, Ca AS $CaSO_4 + 2H_2O$ (TO DIFFERENTIATE FROM QUARTZ, ETCHED WITH DILUTE HF FOR 10 SECONDS QUARTZ REMAINS CLEAR, FELDSPAR BECOMES CLOUDY)
FELDSPAR	ORTHOCLASE $K_2O\text{-}Al_2O_3\text{-}6SiO_2$	COLORLESS			MONOCLINIC NEG.					PRISMS GRANULES	1.53-1.54	0.005		$a:\mathfrak{u}=5°$	2V = 69°-43' 2E = 121°-6'	6-6.5	2.5-2.6	GELATINIZES IN HCl
MUSCOVITE SERICITE	$K_2O\text{-}3Al_2O_3\text{-}6SiO_2\text{-}2H_2O$	COLORLESS			MONOCLINIC NEG.					BIRD'S-EYE MAPLE APPEARANCE, FLAKES LEAVES & SHREDS	1.56-1.60	0.037		$C:\mathfrak{u}$ = 0° to 2°	2V = 30° to 45° 2E = 60° to 90°	1-2	2.8-3.1	NOT SOLUBLE IN HCl & H_2SO_4 BUT READILY IN HF, K as K_2PtCl_6.
RUTILE	TiO_2	YELLOWISH BROWN				TETRAGONAL POS.	↓		↓	GRAINS ACICULAR CRYSTALS	2.50-2.90	0.260	SELDOM NOTICEABLE IN THIN SECTIONS	$C=\mathfrak{c}$		6-6.5	4.2-4.3	NOT SOLUBLE IN ACIDS, FUSED WITH Na_2O_2, Ti as $Rb_2TiF_6 + H_2O$
CHLORITE = CLINOCHLORE	$H_8Mg_5Al_2Si_3O_{18}$	GREENISH				MONOCLINIC POS. SOMETIMES	NEG.		SOMETIMES NEG. RHOMBIC	FLAKES LEAVES	1.59-1.60	0.011	$b:\mathfrak{u}$ = GREEN $\mathfrak{t}$ = YELLOWISH	$C:\mathfrak{c}$ = 2° to 9°	2E = 32°-90°	2-2.5	2.6-2.8	SOLUBLE IN CONCENTRATED H_2SO_4
TOURMALINE	Na Al B Mg Cr Fe Plus SILICATES	REDDISH BROWN TO BLUISH VIOLET						HEXAGONAL NEG.		PRISMS AND ACICULAR CRYSTALS	1.65-1.68	0.032	STRONGEST AT RIGHT ANGLES TO VIBRATION OF LOWER NICHOL			7-7.5	3-3.2	NOT READILY SOLUBLE IN ACIDS FUSES WITH Na_2CO_3
BIOTITE	$(KH)_2\,(MgFe)_2\,(AlFe)_2\,Si_3O_{12}$	GREENISH-BROWN TO RED-BROWN.						MONOCLINIC NEG.		BIRD'S EYE MAPLE APPEARANCE, FLAKES LEAVES & SHREDS	1.56-1.60	0.040	STRONG	$C:\mathfrak{u}$ = 0° to 7°	2E = 0°-50°	2.5-3	2.8-3	SOLUBLE IN CONCENTRATED H_2SO_4, Mg AS $NH_4MgPO_4 + 6H_2O$
CALCITE	$CaCO_3$	COLORLESS GRAY						HEXAGONAL NEG.		GRANULES & CAVITY FILLINGS	1.49-1.66	0.130	WEAK			3	2.6-2.8	READILY SOLUBLE IN HCL & $HC_2H_3O_2$
ANDALUSITE	$Al_2O_3\text{-}SiO_2$	COLORLESS PINK							RHOMBIC NEG.	PRISMS IRREGULAR GRANULES	1.63-1.64	0.011	WEAK	$C=\mathfrak{u}$ $a=\mathfrak{c}$	2V = 85°	7-7.5	3.1-3.2	SOLUBLE ONLY IN HF
RHODOCHROSITE	$MnCO_3$	DARK RED TO BLACK						HEXAGONAL NEG.		FLAKES & GRANULES	1.83	0.010	WEAK			3.5-4.5	3.3-3.6	SOLUBLE IN HCL (CO_2) Mn AS $MnC_2O_4 + 3H_2O$
BARITE	$BaSO_4$	GRAYISH WHITE							RHOMBIC POS.		1.64-1.65	0.010	WEAK		2E = 63°	3-3.5	4.3-4.7	INSOLUBLE IN HCl & HNO_3, FUSED WITH Na_2CO_3, Ba AS $BaSiF_6$

The use of shafts or drifts in the mining of slate is resorted to only where the vein is narrow and vertical or nearly so. The active slate mines of this type are now confined to the Monson district of Maine.

Here each mine demands a slightly different method of treatment, some sinking a shaft and working the roof of the drift to obtain the slate utilizing the waste to build up the floor. Others combine this method with that of sinking the drift floor, taking slate from the floor and working down the bed.

Landing for Blocks from Quarry Cableways enroute to Mill

In each, however, fullest advantage is taken of the seams, making these take the place of a channel cut wherever possible.

Hand drills, machine drills and blasting are used to some extent, depending upon the location of the bed. The use of channeling machines, run by steam or compressed air, which give a smooth surface and provide a means of obtaining regular rectangular blocks, is found in Pennsylvania quarries.

A Modern Mill Shapes Roofing Slate

After the block is cut by the channeling machine, it is necessary to separate it from the floor. Where seams do not occur, horizontal drill holes are projected at the floor of the bench parallel to the slaty cleavage. Black blasting powder exploded in small charges then makes a fracture, separating the mass from the floor. These large masses are subdivided into slabs by splitting along the cleavage plane and can then be hoisted to the surface largest and most up-to-date quarries gasoline engines

From the cableways the blocks are placed on cars at the landing and drawn to the mill or shanty. At the largest and most up-to-date quarries gasoline engines draw the cars directly into the mill. Here the slabs are lifted by electric traveling cranes and placed on saw tables where they are sawed and later "skulped" into convenient size slabs for the slate splitter.

Splitting and Trimming Roofing Slate

The slate splitter, using a flexible chisel and wooden mallet, splits the slab first in the center and continues to subdivide it in the center until slate are obtained of the required thickness. The final pieces are then trimmed to commercial size either on a foot treadle machine or by a mechanical trimmer, and hauled to the storage yards and piled on edge, each pile being made up of slates of the same size. Slates are sometimes punched before piling; at other times just before shipment, and sometimes on the job.

Roofing Slates on the Quarry "Bank"

Possibilities in the Use of Roofing Slate

ALBERT KELSEY, F. A. I. A.

INASMUCH, as architectural training and taste are advancing with giant strides in the United States, every architect appreciates the help he is constantly receiving from well digested and carefully standardized literature issued by well established and reliable associations representing different industries. Not that he wants his work to be standardized—far from it. But he does want reliable information in concise form and new and fresh ideas to help him in the use of the various materials he must incorporate in his designs and specify.

No more convincing evidence of this is to be found, than in turning the leaves of half a dozen publications, issued by as many associations, each pleasingly setting forth in text and pictures, new and proper ways to use their respective products.

The text is often a model of condensation and clearness, while the pictures frequently are as valuable and as beautifully presented as those in large and expensive volumes, that many architects feel they cannot afford.

That the slate industry is represented among these enlightened industries, is gratifying, since the average architect knows far less about the use of slate than about some other products that have long appealed to him because of their well-known architectural possibilities.

Now in this book, I am limited to the subject of roofing slate only, and, therefore, I think first of all of the finest and most famous slate roofs in the world—those of the stately Chateaux in the valley of the Loire. Fortunately, through the kindness of Mrs. C. C. Zantzinger, whose husband is a well-known architect, I am able to reproduce on the next page, a photograph of Azey le Rideau, usually considered the most well-bred of all the Chateaux of the Francois Premier period, a picture that not only shows the type of roof I wish to describe, but one having the charm of the accidental about it.

To be sure, the Chateaux itself, is only suggested, but its impressive gate houses and stately gateway, set off by grand old trees, show off the roofs in the foreground magnificently. Please observe these great gleaming surfaces, as tall as the elevations themselves, but though the roofs are large, it is interesting to note that the smallest size and the thinnest slates (indeed, the cheapest) have been purposely used in order to increase the scale of the roofs and to insure a fine, smooth elegance. It is the smoothness and the fine texture that makes them so distinguished. Note the hip-lines in particular, while the metal cresting has been treated with boldness in order to make the slate all the more delicate and refined in appearance.

In Germany, for centuries, builders have woven designs in slate that are a delight still to every beholder

Turning next to a German illustration, it is interesting to note how slate has been used in this high roof as an edging, much like a piece of braid around the edge of a fashionable cutaway coat, and again how beautifully the slate have been woven in around each of the dormer windows. Here are suggestions and many new possibilities, for it is possible to use up all the small and irregular and inexpensive run-of-the-quarry slate for the field, and the more varied in color and texture, the better; and then large selected slate of one color only for the edges and the frames around the dormer windows.

One illustration, page 79, shows thick roofing slate put to a totally novel use. We see it here, the rough side out, forming a most artistic and durable wainscoting; and it is easy to imagine the delicate blendings of purples, greens and blues that add a note of spice to the interesting roughness and grainings of such picturesque surfaces.

The dozen attractive pencil sketches prepared by Mr. Mecaskey suggest a hundred caprices that the architect may indulge in, if he has the patience and enthusiasm to spice up his work with a few fragments of slate. Architect Andrew Thomas, of New York, has done this frequently with most pleasing results, one of which Mr. Mecaskey shows in his last sketch on page 82.

In the first illustration, with less than a dozen thick pieces of slate, he gives scale and distinction to the most

Formal Entrance to the Chateau of Azey le Rideau

Great gleaming surfaces made up of little slate laid only four or five inches to the weather give elegance to these roofs. It is the smoothness and fine texture that makes these roofs so distinguished

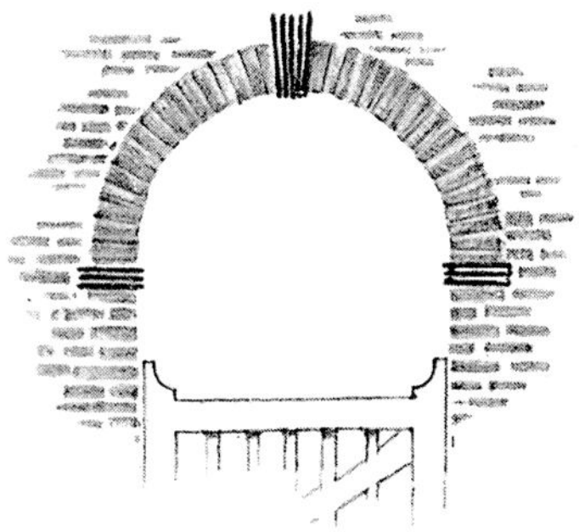

commonplace archway. In the next, he uses slate more boldly for a little shelf over a triple window, and thereby adds a note of refinement to the entire window. In another arch, he separates the local stone by individual slivers of slate, thereby making the arch sparkle, while in the fourth, he suggests how a rich front door mat effect may be produced by using slate on edge as a landing.

His fertile mind has evolved a number of other illustrations of similar uses of slate, which the reader can decipher for himself, though I cannot refrain from commenting upon the happy combination of the curved edge of a great roof just slightly projecting beyond the face of the wall and then the eaves corbeled out to receive the over-hang by means of many layers of slate, all of which produces a change in scale that is altogether charming, when used on an informal and romantic structure, but which would be utterly out of place on the Chateaux of Azey le Rideau.

I am surprised, however, that he has not furnished one illustration showing the use of slate in garden stairways. Nothing adds more to the rustic charm of a garden than steps and stairs built with treads not over 1½″ thick out of rough slate and with risers built up of slate not over ¾″ thick, with their joints well raked out.

Also at the risk of going beyond the limitations of this subject, I wish to say that if a tablet with an inscription is to be built in a wall, there is no other material with so fine a grain that will take the most delicate lettering so well. In support of which, I need only point to the cemeteries of New England, where one still sees countless old tombstones, sometimes 200 years old, on which the inscriptions are as clear as the day on which they were cut.

I understand Architect Blanchard used slate memorial tablets as well as roof and blackboards for Rockville Centre School shown at lower right, page 81. Lower right, page 80, shows unusual wall coping of slate. Has not the Association made a happy selection of roof treatments to intrigue the designer with the possibilities of the use of roofing slate?

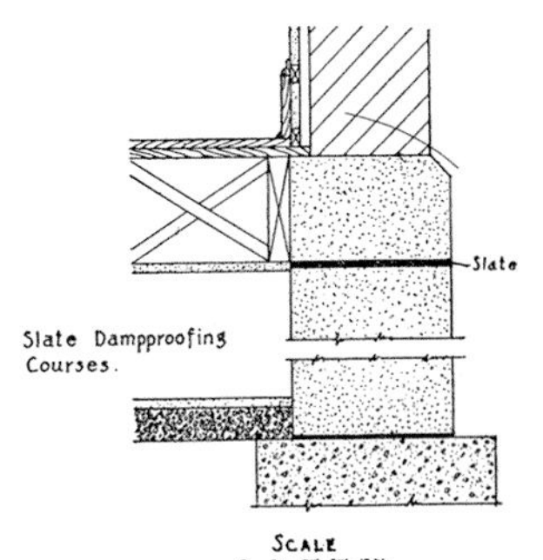

Slate as a dampproofing course has been used for many years. The Building Code of the District of Columbia states as follows:— "A course of slate must be worked into all walls to the full width of the walls, two courses above the surface of the ground and an additional course above footings where there is a cellar or basement."

A few suggestions of the possibilities, both decorative and protective, in the use of roofing slate at the command of the designer

As one approaches the great Bear Mountain Bridge over the Hudson the slate roofs of the two gate houses seem to blend into the natural stone and foliage colors and become an integral part of the whole charming scenic effect. The other pictures show variations in the treatment of the details of slate roofs by which architects have secured unusual and pleasing results

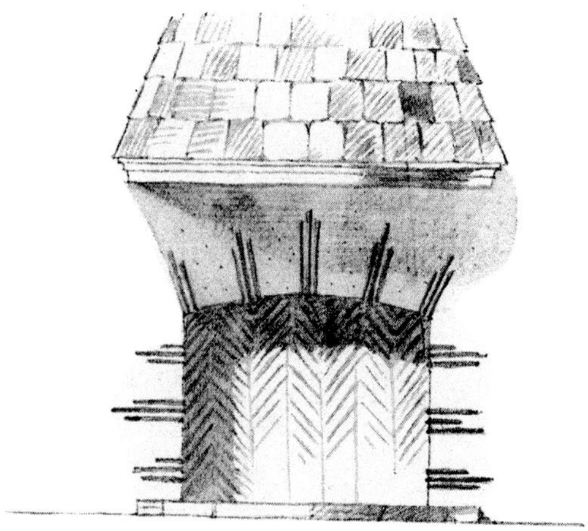

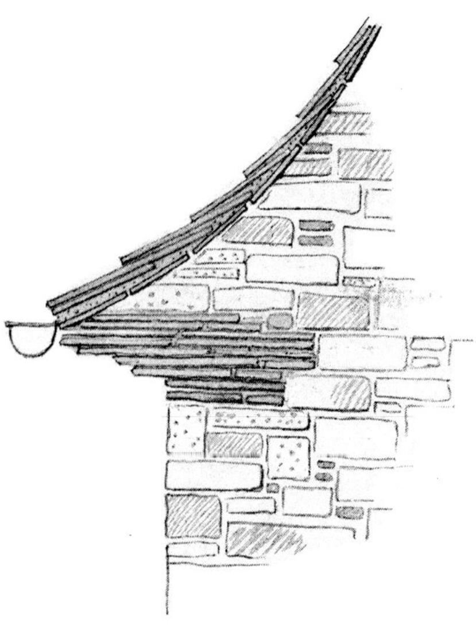

INDEX

NATIONAL SLATE ASSOCIATION

MEMBERS ARE RELIABLE SOURCES OF SUPPLY

Active Members Insignia

This publication, together with recent brochure on "Charm of Slate Floors and Walks" and other informative literature on slate, have been made possible only through the complete co-operation of the members of the Association.

The officers of this Association desire to thank the active or quarry and mill members, a substantial majority of the slate industry, who have so generously contributed of their time and money to make these works available to the public. These members have been imbued at all times with the idea that the public should benefit from a wider knowledge and use of slate and that slate for its many uses may be properly installed to give the service warranted by the inherent qualities of this natural stone.

Associate Members Insignia

In the past, when roofing slates were produced by haphazard and rule of thumb methods, some slates may have found their way on to roofs which present-day operators influenced by results of scientific knowledge and technical investigations would not tolerate. This is the trend of times in all industries for better products and better service from products. The National Slate Association is striving to secure for all users of roofing slate the centuries of sheltering service and high salvage value that this stone will give. Membership in this organization is an evidence of desire and willingness on the part of any quarry to furnish only high-grade, properly selected slate. Users of roofing slate will do well to specify and insist that their requirements come from quarries of members. This is not said with any idea of restricting competition for any new capital or active concerns, for the development of quarries in the worthwhile deposits will be gladly welcomed and encouraged by the Association.

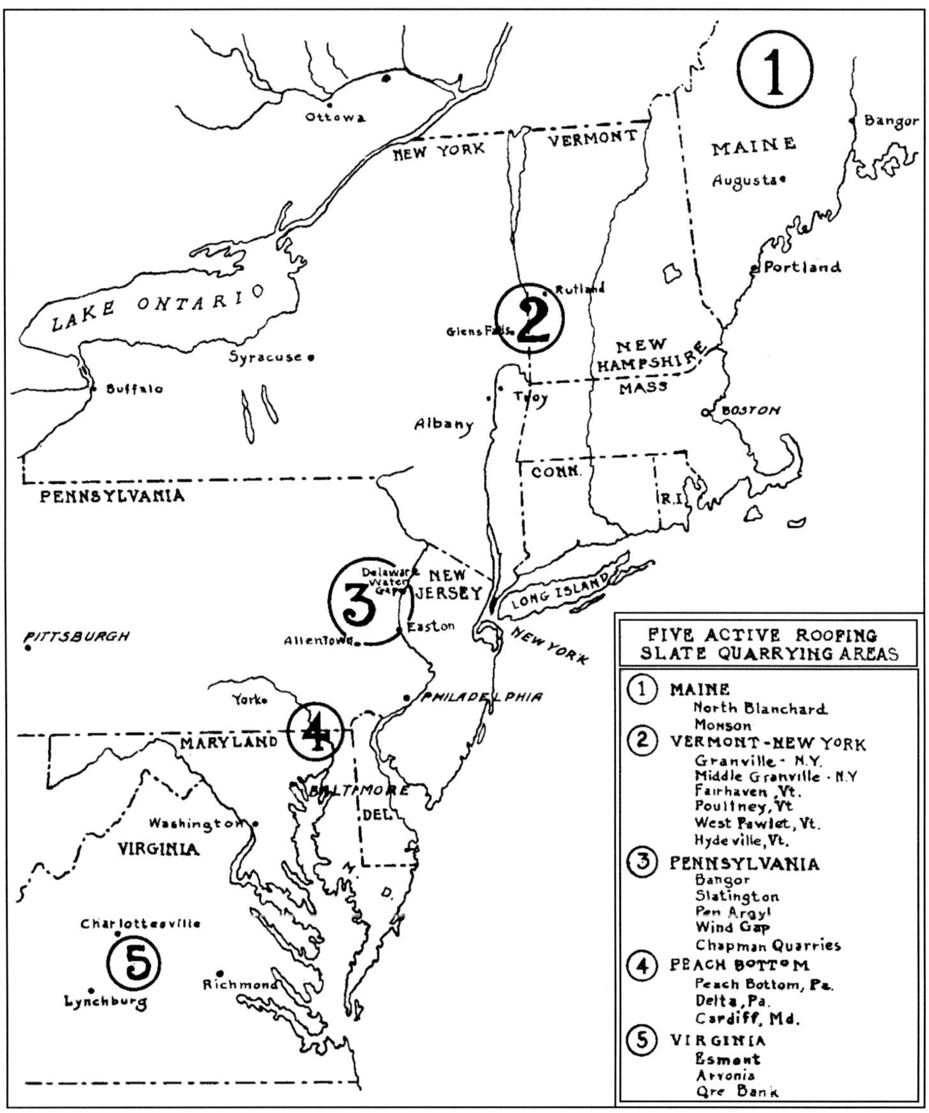

To protect the public the Association stands ready at all times to identify and classify any slate. Recently it has developed that some slates formerly quarried do not come up to present standards of members so that architects and owners will do well to obtain a report from the Association on any slates submitted for a roof from other than quarries of members or deposits not located in the five distinct and active quarrying areas of the United States.

Architects, builders and owners may be assured that they will be dealing with responsible parties if they obtain estimates and information on their slate requirements in their localities from associate members of the National Slate Association. Such slate distributors, roofing and setting contractors are readily identified by their membership insignia displayed on letterheads, bidding sheets, invoices, advertising, etc., or at their offices.

General Guidelines for Natural Quarried Roofing Slate

Slate Roofing Contractors Association of North America, Inc.

2017

These guidelines are included with the permission of the board of directors of the Slate Roofing Contractors Association of North America, Inc. as authorized by unanimous vote of the board at a regular meeting on February 6, 2020.

General Installation Guidelines for Natural Quarried Roofing Slate

Slate Roofing Contractors Association of North America, Inc. 2017

DISCLAIMER

The information contained in the Guidelines is provided for educational purposes only. Persons considering adoption of any practices or standards described in these Guidelines should review all pertinent federal, state, local, and contractual regulations, and abide by any applicable building, safety, and health codes related to construction practices or use of equipment. These Guidelines are not intended to be a source of information regarding safety issues associated with roof systems, procedures, techniques, or products, nor do these Guidelines address required or recommended safety procedures relevant to roofing. Users are urged to take all necessary safety measures as outlined in product safety data sheets, OSHA requirements, state regulations, and local ordinances.

By using the information contain in these Guidelines, the user expressly assumes all risk of loss, harm or injury resulting from the use or misuse of such information. Modification to all or parts of the information included in these Guidelines by the user may affect the results, accuracy, reliability or other aspects of the information. The user also acknowledges and accepts the foregoing limitation of liability and disclaimers and agrees the Slate Roofing Contractors Association of North America Inc. (SRCA) is not responsible for injuries, claims, losses or damages to the user or third parties arising directly or indirectly from the users' use of these Guidelines. Under no circumstances shall the SRCA, its officers, or any of its respective partners, directors, employees, agents, associates, or representatives be liable for any damages, whether direct, indirect, special or consequential damages for lost revenues, lost profits, or otherwise, arising from, or in connection with, this publication, or the materials contained herein.

The SRCA may, from time to time, amend or update these Guidelines. Since various construction, roofing, and safety regulations, codes, standards, and practices are subject to change, these Guidelines may be current only as of the original publication date, and users should consult appropriate sources for the most recent versions of all codes, regulations and standards. The use of these Guidelines is intended to be voluntary, and the SRCA assumes no liability in connection with the use or misuse of these Guidelines or information included therein, or in connection with any modifications to the Guidelines by a user, licensee, instructor, sub-contractor, or trainer, and expressly disclaims any such liability or responsibility

COPYRIGHT AND LICENSE TO REPRODUCE

INTRODUCTION

These Installation Guidelines have been developed and produced for the general public by professional slate roofing contractor members of the Slate Roofing Contractors Association of North America, Inc., a 501c6 non-profit international trade association, via a peer-reviewed, consensus-based committee process taking place over an extended period of time. These Installation Guidelines, if followed, are designed to produce a long-lasting, weathertight slate roof for most basic applications. Refer to the SRCA Section 07310 "Slate Shingles" for architectural specifications.

PART 1 — GENERAL PROJECT PREPARATION AND EXECUTION

1.1. CONTRACTORS

A. It is recommended that prospective contractors have foremen or supervisors who have a minimum of five years of professional experience in installing natural quarried slate roofing and who can provide a resume listing successful completion of slate roof installations of similar size and scope.

B. Contractors shall furnish all insurance, permits, labor, materials, equipment, apparatus, tools, transportation and services necessary for, and incidental to, the proper installation and completion of the slate roof. This work may include removing and properly disposing off-site of existing slate roofing or other shingles, if any; installing underlayment; installing new sheet metal flashings, and installing new or salvaged roofing slate to cover the entire existing slate roof area, leaving a very long term, damage resistant, weatherproof roof.

C. Contractors shall use workmen who are trained and experienced in laying slate, installing metal flashing, and all other skills needed to satisfactorily complete the project as specified, or use workmen who are under the full-time supervision of a foreman or supervisor with such training and experience.

D. Contractors shall use workmen familiar with the use of slate hammers for punching and nailing slate shingles, slate rippers for removing slates already installed, slate cutters for trimming and cutting slate shingles, slater's stakes used with slate hammers that have a cutting shank, roof brackets and scaffolding for staging the roof, and hook ladders for accessing areas of the roof not staged.

1.2. CONTRACT DOCUMENTS

A. Contractors will provide contract documents that include detailed specifications for all materials and installation styles, including, but not limited to slate types, thicknesses, colors, and origins; nail types and lengths; underlayment types and weights; headlaps; ridge style; valley style; flashing materials, including weights, types and sizes of metals.

B. Contract documents shall also include a detailed warranty (see Section 1.8). It is recommended that this SRCA General Roofing Installation Guidelines for Natural Quarried Slate document be attached to the contract as an addendum.

1.3. MATERIALS DELIVERY AND STORAGE

A. Slate roofing shingles shall be delivered on manufacturer's skids with intact labels indicating source whenever possible.

B. Rolled underlayments shall be stored on end and manufacturer's recommendations for job-site storage and protection shall be followed.

C. Materials shall not be stored on existing fire escapes, in areas that may obstruct the ingress or egress of essential doorways or walkways, or on low-slope roofs or other roof slopes where the weight of the materials may damage the roof structure.

1.4 CODES AND REGULATIONS

A. Contractor shall comply with all federal, state, local and contractual regulations regarding specific installations, and abide by applicable building, safety and health codes related to construction practices or use of equipment.

B. Contractor shall comply with the regulations of local governing Historical Societies and the National Registry of Historic Properties, when applicable.

1.5 PROTECTION OF ROOF SURFACES

A. Workers shall not damage slates by walking on them.

B. The roof shall be properly staged to allow safe work surfaces, such as with roof brackets and scaffold-grade planks, that prevent unnecessary foot traffic on the slates.

C. Where foot traffic is unavoidable, roof ladders, hook ladders, chicken ladders, foam pads or other such devices should be used to protect the slates.

1.6 INSURANCE

A. Contractor shall carry Liability Insurance and Workers Compensation Insurance according to state requirements.

B. Contractor will provide, upon request, Certificates of Insurance to Owner prior to the execution of any work.

1.7 CLEAN UP

A. Tools, equipment, surplus materials, slate scraps, and debris resulting from the slate roof installation shall be organized and cleaned up, or removed and properly disposed of, on a daily basis.

B. Gutters and roof areas will be cleaned of debris at the end of each work day and upon completion of the work.

C. Dust and dirt may infiltrate into the attic space during installation or removal of roofing slate. Owner shall be advised to remove any valuable items from the attic space and/or to cover such items with plastic, tarps, or other suitable covering prior to the commencement of any work.

1.8 WARRANTY

A. Contractor shall warranty a new slate roof installation, covering both material and labor for defects that may occur, for a minimum period of five years.

B. Owner shall visually inspect the Work on an annual basis and report to the contractor any observed defects that are covered by the warranty.

C. The contractor cannot be held responsible for damage caused by other persons or by unusual and damaging weather events such as large hail, strong winds, lightning, flooding, earthquakes, excessive snow and ice buildup, or other "acts of God."

D. Contractor's warranty terms shall be clearly detailed in the contract documents.

E. An extended service maintenance agreement is recommended beyond the warranty period.

1.9 EXTRA MATERIAL

A. Contractor shall provide for the Owner two percent of field slates used in the Work as maintenance stock for future repairs.

B. The maintenance stock is to be placed in storage in a location at the work site to be determined by the Owner.

PART 2 — ROOFING SLATE

2.1 PROCUREMENT

A. Procure new roofing slates from known sources so that additional matching supplies can be obtained if needed. A single quarry source is recommended.

B. Slates shall be supplied by manufacturers that are experienced in the production of roofing slate and that quarry or mine high-quality rock specifically selected for roofing grade applications. Provide manufacturer's warranty in writing. Substitution of slates with slates from other sources shall not be permitted without written approval from the Owner or architect.

C. Slates shall be identified and labeled according to the quarry source and location. For example, foreign slates shall not be misrepresented as domestic slates.

D. When ordering field slates, whenever possible, also procure the starter slates from the same manufacturer, punched for nail holes to allow the starter slates to be laid back side out. The starter slates should be the same width as the field slates (when uniform width slates are used), and long enough to allow for a minimum headlap of three inches on the starter course.

E. Refer to http://slateroofers.org/sources_new_slate.html for sources of new roofing slate.

2.2 QUALITY CONTROL

A. Slates shall conform to ASTM C 406 and shall be Grade S1 (minimum 575 lbf breaking load[1], maximum 0.25% absorption[2], and maximum 0.002 inches depth of softening[3]).

> *[[1]ASTM C120, Test Methods of Flexure Testing of Slate (Breaking Load, Modulus of Rupture, Modulus of Elasticity); [2]ASTM C 121, Test Method for Water Absorption of Slate; [3]ASTM C 217, Test Method for Weather Resistance of Slate]*

B. All slate shall be hard, dense, sound rock of natural cleft with chamfered (beveled) edges. No broken or cracked slate shall be used, although broken slates may be cut into smaller, unbroken pieces.

C. Slates up to 3/8" in thickness shall be punched for nail holes; slates over 3/8" in thickness may be drilled and countersunk, for a minimum of two nail holes each. The holes punched in the slates shall be the correct diameter to provide a snug fit for the shank of the roofing nails. Slates shall be punched back to front (except starter slates, which are punched front to back). Slates shall be punched on the thinner end if there is a variation in the thickness along the length of the slate.

D. Rectangular slates with broken corners on the exposed end shall be rejected if a corner is broken off greater than 1.5 inches in either direction, although such slates may be used for cutting into smaller pieces.

E. Curvature of slates shall not exceed 1/8 inch over a distance of 12 inches. Curved slates shall be trimmed and punched to permit them to be laid with the convex side facing up.

F. Defects in slates such as "knuckles," "knots," "knurls" and "cramps" which protrude above the surface of the slates shall be positioned such that they remain on the exposed top surface of the slate after installation. Knots, knurls and cramps shall not be permitted on the back or covered portions of the slates unless removed by grinding beforehand. A slate shall be rejected if a surface defect adversely affects the proper laying of the slate.

G. Slates shall be free of pyrite inclusions that can visibly leach rust stains onto the roof.

H. Slates shall not have carbon-bearing bands known as "ribbons" as these are considered defects that undermine the longevity of the slate.

I. Nail holes are to be positioned no more than 1.5" in from the side edges of the slate. Nail holes must be positioned approximately 2/3 the distance from the bottom of the slate when using standard 3" headlap. The top of the underlying slate shall not be penetrated by the slating nails.

2.3 INSTALLATION STYLES

A. Slate installation styles may include, but are not limited to:

1. A "standard" pattern where all of the slates are the same length and width, although these can include slates of differing colors and shapes;

2. A "random width" style where the slates are the same length, but differing widths;

3. A "textural" style where thicknesses, textures, colors, lengths, widths and types of slate may be blended for architectural effect;

4. A "graduated" style where thicker or longer slates are installed near the eaves and the slates gradually decrease in size as they near the top of the roof;

5. A "staggered butt" style where slates of varying lengths are installed in such a manner that the exposed butts are staggered.

B. When mixed, graduated, or textural slate styles are to be installed, or slates with mixed colors, sizes, types and/or thicknesses, the quantities of the various sizes and types of slates and the layout patterns of the slates are to be clearly specified in the contract documents.

C. If necessary, build mockups to demonstrate aesthetic effects and to set quality standards.

1. Use materials specified for the project in the mockup.

2. Retain mockup for duration of the project.

3. Approved mockups may become part of the completed work.

PART 3 — GENERAL MATERIALS

3.1 ROOFING UNDERLAYMENT

A. Cover surfaces to be slated with roofing underlayment to weather in the building, when needed.

B. Roofing underlayment shall, at a minimum, comply with ASTM D 226 asphalt-saturated organic felt, Type II, No. 30, unperforated.

C. Felt shall be installed horizontally with sections overlapped toward eaves or drains by a minimum of two inches and at ends by a minimum of six inches. The felt shall overlap hips and ridges by approximately 12 inches. All felt shall be preserved unbroken, tight, and whole.

D. Felt shall be secured with minimum 1" electro-galvanized roofing nails along laps, ends, and in the field as necessary to properly hold the felt in place and to protect the building from water infiltration until covered with slate.

E. The maximum length of exposure for felts prior to slating shall be one month. When the felt must be left for long periods before the slates can be installed, the exposed nail heads are to be skimmed over with a thin layer of trowel grade roof mastic to prevent leakage around the nail heads.

F. When self-adhesive underlayment is used along eaves or elsewhere, it shall be covered with felt.

G. On slopes from 4:12 to 12:12, a half-lapped double layer of No. 30 felt is recommended when the felt is not going to be slated over immediately. Slopes over 12:12 may use a single layer of felt as may any slope that is to be slated within a few days.

H. Additional underlayment may also be used in ice-dam prone areas, such as Type II felt installed on top of the existing felt, with a layer of trowel-grade roof mastic spread evenly underneath the additional felt layer.

3.2 NAILS

A. Nails shall be solid copper, smooth-shank roofing nails, minimum 11 gauge, minimum 1.25" length. Copper nails 2.5" or longer shall be minimum 10 gauge. Alternatively, Type 304 smooth-shank stainless steel roofing nails can be used, not less than 1.25" long.

B. Hot-dipped galvanized smooth-shank roofing nails may be utilized when specified.

C. Nail length is to be approximately twice the thickness of the slates plus one inch. Nails are to fully embed into the roof decking material without more than 1/4" nail length being exposed on the underside of the roof decking. When the underside of the roof decking is exposed, such as at overhanging eaves, the nails shall be long enough to penetrate the roof decking, but not so long that they may be visibly driven through.

D. Recommended nail lengths are as follows when 1" or thicker roof deck is utilized:

1. 3/16"-1/4" thick slates are to be fastened with 1.5" nails.
2. 3/8" thick slates are to be fastened with 1.75" nails.
3. 1/2" thick slates are to be fastened with 2" nails.
4. 3/4" thick slates are to be fastened with 2.5" nails.
5. 1" thick slates are to be fastened with 3" nails.

3.3 FLASHING

A. Flashing shall be minimum 16 ounce copper conforming to ASTM B 370, minimum 28 gauge stainless steel, or minimum 4 lb. sheet lead. Twenty ounce copper flashing is recommended.

B. Flashing shall be installed where there are roof plane intersections, where the roof abuts walls, parapets, dormers and chimneys, or where there are roof penetrations.

C. All flashings and fasteners are to be galvanically compatible metals.

D. Additional flashing guidelines are available from SMACNA's "Architectural Sheet Metal Manual" and the publication "Copper and Common Sense" by Revere Copper Products.

3.4 MASTIC

A. Roof mastic shall be non-asbestos fibered asphalt cement complying with ASTM D 4586.

B. Roof mastic shall be designed for trowel application.

3.5 SLATE HOOKS

A. Slate hooks shall be minimum three inches long, solid copper or stainless steel.

B. Standard slate hooks are for use with slate of commercial standard thickness only. Custom slate hooks may be fabricated for use with thicker slates.

3.6 CANTS

A. Wood cants for installing underneath the starter course of slate shingles shall be minimum 1/4 inch by 1 inch lath or other solid, glue-free wood. Alternatively, minimum 16 ounce copper or 28 gauge stainless steel edging with a built-in cant may be used.

B. When installing standard thickness slates (3/16" to 1/4" thick) the cant should be approximately 1/4" to 3/8". Thicker slates will require a thicker cant.

3.7 ROOF DECKING

A. The roof deck shall be a minimum of 3/4" thick wood. Solid, glue-free wood is recommended.

B. Nailable concrete and gypsum concrete may also be suitable roof decking materials.

C. Minimum 3/4" glue-free slating lath or skip sheathing can be spaced on rafters as a nailing substrate.

D. Minimum 3/4" glue-free boards, slating lath or skip sheathing can be installed over glued or laminated roof decks to provide a suitable nailing substrate.

E. Surfaces to which the roof slates are to be applied shall be in a suitable condition or shall be repaired to a condition satisfactory for slating. All surfaces to be slated shall be swept clean of any debris.

PART 4 — SLATE EXECUTION

4.1 FASTENERS

A. All standard slates shall be fastened with minimum two roofing nails fastened above the head of the underlying slate and as far from the center of the slate as is practical.

B. Larger, heavier slates may need four nails per slate.

C. Screws shall not be used when fastening slates.

D. Slates overlapping sheet metal shall have the nails placed so as to not puncture the metal.

E. Exposed nail heads are not permissible except where unavoidable. Any exposed nail heads shall be sealed with gaskets or approved sealants. The application of slate dust to cover exposed sealants is recommended.

F. Nails shall not be driven in so far as to produce an excessive strain on the slates, and shall instead be driven to a depth such that the nail heads lie within the counter-sunk nail hole and do not rub excessively against the overlying slates.

G. Use of pneumatic or electric nail guns to install slate shingles shall not be permitted.

H. Slates shall not be bedded in roof mastic or other adhesives except where absolutely necessary such as at exposed edges in high-wind areas.

I. Ridges and hips shall be installed without exposed fasteners.

J. If a slating nail is installed in a crack or hole in the sheathing, it shall be renailed properly.

4.2 HEADLAP

A. All standard field slates shall be installed with a minimum 3" head lap when the roof slope is 8:12 up to 20:12.

B. Less than an 8:12 slope down to 4:12, the slates shall be installed with a minimum 4" headlap.

C. Installing roofing slates on slopes less than 4:12 is not recommended.

D. On slopes 20:12 or greater, slates may be installed with a 2" headlap.

E. Headlaps may be increased at ice-dam prone or poor drainage areas.

4.3 EAVES SLATES, EDGE SLATES AND STARTER COURSES

A. Eave slates shall be laid to provide a minimum 1.5" projection beyond the furthest extent of the fascia, cornice, crown molding, metal drip edge, trim, or other construction material at the eaves.

B. Rake edge (gable end) slates shall extend 1" beyond the furthest extent of the gable trim, fascia or edge.

C. Slates at the eaves shall be doubled by first installing a slate starter or under-eave course installed back side up (chamfered side down). The first course of slate shall be laid over the starter course so that the drip edges of both courses align flush. The first course of slates shall break side-butt joints with the starter course side-butt joints by not less than 3". The second course of slates must overlap the starter course by a minimum of 3" and not less than the general headlap of the field slates along the eaves [see Figure 1].

D. The starter course of slates is to be canted to allow the starter slates to be tilted to the same angle as the field slates.

4.4 SLATE INSTALLATION

A. Slate shall be installed starting at the bottom or eaves and proceeding toward the ridge or top.

B. All slates will be installed following chalk lines marking the top edge of each course of slates, whenever possible.

C. When supplied on pallets, slates are not to be used from one pallet at a time, but are to be used from all pallets simultaneously in order to blend the various pallets uniformly on the roof.

D. Slate side-butt joints shall be positioned as near the mid-point of the underlying slates as possible, and not less than 3" from the underlying side-butts. Each slate course shall break butt-joints laterally by a minimum of 3", if possible, with the underlying or overlying courses.

E. When installed, slates shall be laid side-to-side with a maximum 1/8" gap between slates, on average, unless otherwise specified.

F. Slate will be neatly fitted around any pipes, ventilators, and other roof penetrations.

G. Slates are to be cut from the back side in order to preserve the chamfered edge on the front exposed surface. Use of grinders, saws, or other mechanical means to cut and trim roofing slates shall not be permitted unless the slates maintain a chamfered appearance along the exposed sawn edges. At the apex of slated hips and ridges, and at the juncture of a closed valley, the chamfered edge may, as an option, be reversed to create a tighter fitting joint.

H. Slates along valleys shall be cut in neat and straight lines. Valley slates are to be cut on the back side of the slate to maintain a chamfered slate edge when along an open valley.

I. Contractor shall visually and manually inspect the slates when roof brackets are removed to make sure no slates were broken by the roof scaffolding. Upon completion, all slate shall be sound, unbroken, un-cracked, whole and clean, showing no exposed roof cement.

J. Individual slates that must be installed in the field of the roof after the installation is complete, such as where a roof bracket had been removed or where a repair has been made, shall be installed using stainless steel or copper slate hooks or the "nail and hidden bib" installation method where standard nailing is not possible. The top of bib flashings shall extend to the top of the head of the underlying slate.

K. Contractors shall keep the building weatherproof and make every reasonable attempt to complete the project on schedule. Work is to be scheduled when weather conditions allow the work to be performed according to manufacturer's recommendations and in compliance with warranty requirements.

L. Temporary roof coverings, tarps, or other materials shall remain available onsite in the event that the work must be ceased suddenly due to weather conditions and the roof needs to be weathered in until such time as the work may be continued.

PART 6 — REFERENCE INFORMATION

6.1 ASTM REFERENCES

ASTM A167 — Type 304 Terne coated stainless steel, 0.015 inch (0.38 mm) thick stainless steel core material, coated with 0.092 lb/sq ft (450 g/sq m) Terne alloy on both sides

ASTM A666 — Type 304 stainless steel, soft annealed 2D finish (unless harder temper is required for forming or performance), 0.0156 inch (0.4 mm) thick

ASTM B101 — Specification for Lead-Coated Copper Sheet and Strip for Building Construction

ASTM B370 — Specification for Copper Sheet and Strip for Building Construction

ASTM B749 — Specification for Lead and Lead Alloy Strip, Sheet, and Plate Products

ASTM C406 — Specification for Roofing Slate

ASTM C920 — Specification for Elastomeric Joint Sealants

ASTM C1311 — Specification for Solvent Release Sealants

ASTM D312 — Specification for Asphalt Used in Roofing

ASTM D1079 — Standard Terminology Relating to Roofing, Waterproofing, and Bituminous Materials

ASTM D1970 — Specification for Self-Adhering Polymer Modified Bituminous Sheet Materials Used as Steep Roofing Underlayment for Ice Dam Protection

ASTM D2626 — Specification for Asphalt-Saturated and Coated Organic Felt Base Roofing Sheet

ASTM D2822 Specification for Asphalt Roof Cement, Asbestos-Containing

ASTM D3019 — Specification for Lap Cement Used with Asphalt Roll Roofing, Non Fibered, Asbestos, Fibered, and Non Asbestos Fibered

ASTM D4022 — Specification for Coal Tar Roof Cement, Asbestos Containing

ASTM D4586 — Specification for Asphalt Roof Cement, Asbestos-Free.

ASTM D4869 — Specification for Asphalt-Saturated Organic Felt Underlayment for Steep Slope Roofing.

ASTM F1667 — Specification Standard for Driven Fasteners

6.2 SLATE GRADES

A. ASTM C 406 Grade S1: Expected service life in excess of 75 years.

B. ASTM C 406 Grade S2: Expected service life 40-75 years.

C. ASTM C 406 Grade S3: Expected service life 20-40 years.

6.3 SLATE THICKNESSES, SIZES AND COLORS

A. Thicknesses

1. Standards; Nominal 3/16 inch (5 mm) to 1/4 inch (7 mm)

2. Quarters; Nominal 1/4 inch (7 mm) to 3/8 inch (10 mm)

3. Heavies; Nominal 3/8 inch (10 mm) to 1/2 inch (13 mm)

4. Extra Heavies; Nominal 1/2 inch (13 mm) to 3/4 inch (19 mm)

B. Standard Slate Sizes

1. 24"L X 14"W (610 mm X 356mm)

2. 24"L X 12"W (610 mm X 305 mm)

3. 22"L X 12"W (559 mm X 305 mm)

4. 22"L X 11"W (559 mm X 279 mm)

5. 20"L X 14"W (508 mm X 356 mm)

6. 20"L X 12"W (508 mm X 305 mm)

7. 20"L X 11"W (508 mm X 279 mm)

8. 20"L X 10"W (508 mm X 254 mm)

9. 18"L X 14"W (457 mm X 356 mm)

10. 18"L X 12"W (457 mm X 305 mm)

11. 18"L X 11"W (457 mm X 279 mm)

12. 18"L X 10"W (457 mm X 254 mm)

13. 18"L X 9"W (457 mm X 229 mm)

14. 16"L X 14"W (406 mm X 356 mm)

15. 16"L X 12"W (406 mm X 305 mm)

16. 16"L X 11"W (406 mm X 279 mm)

17. 16"L X 10"W (406 mm X 254 mm)

18. 16"L X 9"W (406 mm X 229 mm)

19. 16"L X 8"W (406 mm X 203 mm)

20. 14"L X 10" W (356 mm X 254 mm)

21. 14"L X 9"W (356 mm X 229 mm)

22. 14"L X 8"W (356 mm X 203 mm)

23. 14"L X 7"W (356 mm X 178 mm)

24. 12"L X 10"W (305 mm X 254 mm)

25. 12"L X 9"W (305 mm X 229 mm)

26. 12"L X 8"W (305 mm X 203 mm)

27. 12"L X 7"W (305 mm X 178 mm)

28. 12"L X 6"W (305 mm X 152 mm)

C. North American Roofing Slate Colors

1. Unfading Black
2. Semi-Weathering Gray/Black
3. Unfading Gray
4. Semi-Weathering Gray
5. Unfading Purple
6. Semi-Weathering Purple
7. Unfading Mottled Green and Purple
8. Unfading Green
9. Semi-Weathering Green (Sea Green, Gray/Green)
10. Unfading Red

6.4. SLATE EXPOSURES AND HEADLAPS

When Using 3" and 4" Headlaps (showing number of slates per square)

Slate Size (in.)	Exposure (3" H.L.)	Slates/Square	Exp. (4" Headlap)	Slates/Square
6x10	3 1/2"	686	3"	800
7x10	3 1/2"	588	3"	686
8x10	3 1/2"	514	3"	600
6x12	4 1/2"	533	4"	600
7x12	4 1/2"	457	4"	515

Slate Size (in.)	Exposure (3" H.L.)	Slates/Square	Exp. (4" Headlap)	Slates/Square
8x12	4 1/2"	400	4"	450
9x12	4 1/2"	355	4"	400
10x12	4 1/2"	320	4"	360
7x14	5 1/2"	374	5"	412
8x14	5 1/2"	327	5"	360
9x14	5 1/2"	290	5"	320
10x14	5 1/2"	261	5"	288
12x14	5 1/2"	218	5"	240
8x16	6 1/2"	277	6"	300
9x16	6 1/2"	246	6"	256
10x16	6 1/2"	222	6"	230
12x16	6 1/2"	185	6"	192
9x18	7 1/2"	213	7"	221
10x18	7 1/2"	192	7"	199
11x18	7 1/2"	175	7"	187
12x18	7 1/2"	160	7"	171
10x20	8 1/2"	170	8"	180
11x20	8 1/2"	154	8"	164
12x20	8 1/2"	141	8"	150
14x20	8 1/2"	121	8"	129
11x22	9 1/2"	138	9"	146
12x22	9 1/2"	126	9"	134
14x22	9 1/2"	109	9"	115
12x24	10 1/2"	114	10"	120
14x24	10 1/2"	98	10"	103

Fig. 1

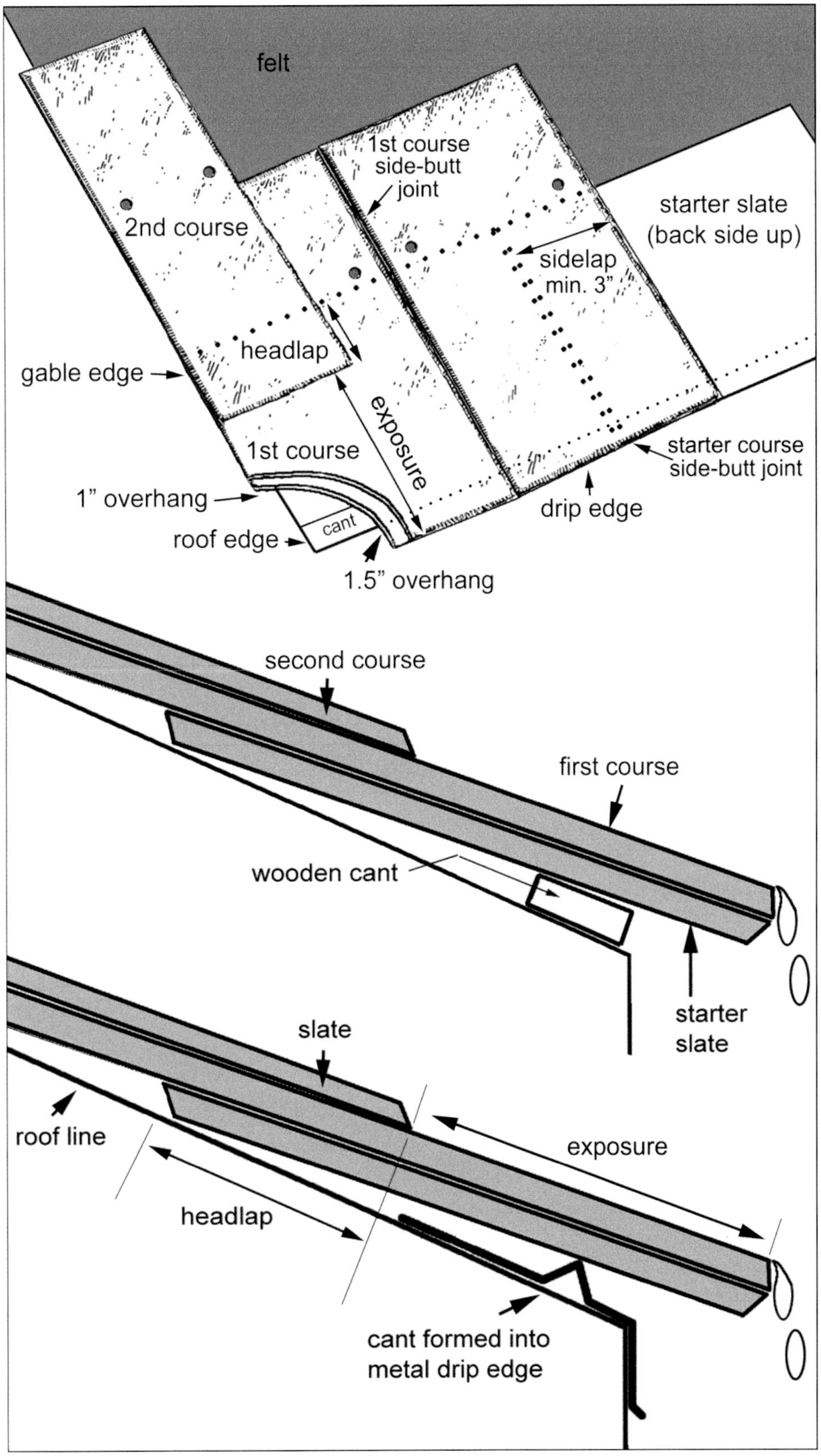

General Installation Guidelines for Natural Quarried Roofing Slate

Slate Roofing Contractors Association of North America, Inc. 2017

Slate Roof Repair, Restoration, and Maintenance Guidelines

Slate Roofing Contractors Association of North America, Inc.

Version 1.1, September 2017

SlateRoofers.org

DISCLAIMER

The information contained in the Guidelines is provided for educational purposes only. Persons considering adoption of any practices or standards described in these Guidelines should review all pertinent federal, state, local, and contractual regulations, and abide by any applicable building, safety, and health codes related to construction practices or use of equipment. These Guidelines are not intended to be a source of information regarding safety issues associated with roof systems, procedures, techniques, or products, nor do these Guidelines address required or recommended safety procedures relevant to roofing. Users are urged to take all necessary safety measures as outlined in product safety data sheets, OSHA requirements, state regulations, and local ordinances.

By using the information contain in these Guidelines, the user expressly assumes all risk of loss, harm or injury resulting from the use or misuse of such information. Modification to all or parts of the information included in these Guidelines by the user may affect the results, accuracy, reliability or other aspects of the information. The user also acknowledges and accepts the foregoing limitation of liability and disclaimers and agrees the Slate Roofing Contractors Association of North America, Inc. (SRCA) is not responsible for injuries, claims, losses or damages to the user or third parties arising directly or indirectly from the users' use of these Guidelines. Under no circumstances shall the SRCA, its officers, or any of its respective partners, directors, employees, agents, associates, or representatives be liable for any damages, whether direct, indirect, special or consequential damages for lost revenues, lost profits, or otherwise, arising from, or in connection with, this publication, or the materials contained herein.

The SRCA may, from time to time, amend or update these Guidelines. Since various construction, roofing, and safety regulations, codes, standards, and practices are subject to change, these SRCA Guidelines may be current only as of the original publication date, and users should consult appropriate sources for the most recent versions of all codes, regulations and standards. The use of the Guidelines is intended to be voluntary, and the SRCA assumes no liability in connection with the use or misuse of these Guidelines or information included therein, or in connection with any modifications to the Guidelines by a user, licensee, instructor, sub-contractor, or trainer, and expressly disclaims any such liability or responsibility

COPYRIGHT AND LICENSE TO REPRODUCE

INTRODUCTION:

These Guidelines have been developed and produced for the general public by professional slate roofing contractor members of the Slate Roofing Contractors Association of North America, Inc., a 501c6 non-profit international trade association, via a peer-reviewed, consensus-based committee process taking place over an extended period of time. These Guidelines, if followed, are designed to produce a long-lasting weathertight slate roof for most basic applications.

PART 1 — ROOF ASSESSMENT

1.1 EVALUATION

A. An evaluation must be conducted by a slate roofer with a demonstrated competency and expertise specifically in repair, restoration, and maintenance of natural quarried slate roofs, not just expertise in new slate roof installation or estimating, or expertise in the repair and restoration of other types of roofs.

B. A roof evaluation should first include an examination of the building interior and underside of the roof, whenever possible, to determine the type and condition of the roof decking, and to identify signs of water penetration or leakage.

C. A thorough evaluation requires accessing the roof and closely examining the conditions of the slates, substrate, fasteners, flashings, and penetrations.

D. A correctly installed roof that has slates with a reasonable remaining longevity (for example, at least fifteen years), is a candidate for restoration.

E. In some cases, full replacement of all or a section of roof will be part of the restoration process.

F. The type(s) of slate must be determined and the remaining longevity estimated.

1. Determine the type, size, shape, thickness, color, weathering characteristics, and texture of the slate.

2. Has the roof been well-maintained, or abused and neglected? Have past repairs been correctly executed? If so, repaired areas should appear very similar or the same as the original roof.

3. No brush-grade or trowel-grade coatings should be covering the slates.

G. Identify the substrate material and condition. Is it solid wood, laminated or glued wood, nailable concrete, nailable gypsum, or metal?

H. Identify the type and condition of the fasteners. If they are nails, what type and length?

Smooth shank or ring-shank? Copper, stainless steel, hot-dipped galvanized, electrogalvanized, cut steel nails, slate hooks, etc.?

I. Determine the existing headlaps and sidelaps.

1.2 CONDITIONS

A. Examine the field slates, starter slates, ridge and/or hip slates, valley slates, etc. How many slates need replaced or repaired?

B. Examine the valley metal, chimney flashings, step flashings, cap or counter flashings, ridge and hip metal, pipe vent flashings, parapet wall flashings, turret and/or tower flashings, etc. What flashings need maintenance work, repair, or replacement? Refer to *Part 6 - Flashings* for additional information on evaluating the conditions of individual types of flashing.

C. When the valley flashing condition cannot be observed, such as on a closed valley, then overlying slates must be removed to expose the flashing. The lower flashings are typically more deteriorated than the higher flashings because they carry more water, so inspect the lower flashings first.

D. What other repairs are needed? Are there flat-lock, soldered metal roofs adjacent to the slate roofs that need replaced? Are there built-in gutter systems? Exterior gutters? Skylights? Chimney or masonry pointing? Snow retention systems?

PART 2 — GENERAL PROJECT PREPARATION AND EXECUTION

2.1. CONTRACTORS

A. It is recommended that prospective contractors have a minimum of five years of professional experience in the repair and restoration of natural slate roofing, and can provide a resume' demonstrating the successful completion of slate roof repair and restoration projects of a similar size and scope.

B. Contractors shall furnish all insurance, permits, labor, materials, equipment, apparatus, tools, transportation and services necessary for, and incidental to, the repair and restoration of the slate roof, unless otherwise agreed upon.

C. Contractors shall use workmen who are trained and experienced in slate roof repair and restoration, removing and re-installing metal flashing, and all other skills needed to satisfactorily complete the project as specified, or use workmen who are under the full-time supervision of a foreman or supervisor with such training and experience.

D. Contractors shall use workmen familiar with the use of slate hammers for punching and nailing slate shingles, slate rippers for removing slates already installed, slate cutters for trimming and cutting slate shingles, slaters stakes used with slate hammers that have a cutting shank, roof brackets and scaffolding for staging the roof, and hook ladders for accessing areas of the roof not staged.

2.2. CONTRACT DOCUMENTS

A. Contractors shall provide contract documents that include a detailed scope of work and specifications for all materials.

B. Contingency fees or unit prices may be included in the contract for unforeseen work such as roof decking replacement, or replacement of additional slates not identified in the original assessment. A contingency fee allowance may also consist of a credit, where applicable.

C. Contract documents shall include a detailed warranty, when required.

2.3 CODES AND REGULATIONS

A. Contractor shall comply with all federal, state, local and contractual regulations, and abide by applicable building, safety and health codes related to construction practices or use of equipment.

B. Contractor shall comply with the regulations of local governing Historical Societies and the National Registry of Historic Properties, when applicable.

2.4 PROTECTION OF ROOF SURFACES

A. Workers shall not damage slates by walking on them.

B. The roof shall be properly staged to allow safe work surfaces, such as scaffold-grade planks, that prevent unnecessary foot traffic on the slates.

C. Roof ladders, hook ladders, chicken ladders, lifts, or other such devices, shall be used to protect the roof surfaces from foot traffic.

2.5 INSURANCE

A. Contractor shall carry Liability Insurance and Workers Compensation Insurance per state requirements.

B. Contractor will provide, upon request, Certificates of Insurance to Owner prior to the execution of any work.

2.6 CLEAN UP

A. Tools, equipment, surplus materials, slate scraps, and debris resulting from the work shall be organized and cleaned up, or removed and properly disposed of, on a daily basis.

B. Gutters and roof areas will be cleaned of debris at the end of each work day and upon completion of the work.

C. Dust and dirt may infiltrate into the attic space during installation or removal of roofing slate. Owner shall be advised to remove any valuable items from the attic space and/or to cover such items with plastic, tarps, or other suitable covering prior to the commencement of any work.

2.7 WARRANTY

A. Contractor's warranty terms, when required, shall be clearly detailed in the contract documents.

B. The contractor cannot be held responsible for damage caused by other persons or by unusual and damaging weather events such as large hail, strong winds, lightning, flooding, earthquakes, excessive snow and ice buildup, or other "acts of God."

C. An extended service maintenance agreement is recommended beyond the warranty period.

PART 3 — MAINTENANCE

3.1 GENERAL MAINTENANCE

A. General maintenance on slate roofs includes:

1. Replacement of any broken, missing or defective slates;
2. Removal and replacement of existing faulty repair work;
3. Painting of flashing metal, traditionally with Tinner's Red or Tinner's Green paint;
4. Emergency repairs.

B. Slate repair as well as removal and replacement of faulty repair work is discussed in Part 5.

3.2 PAINTING OF FLASHING

Old ferrous-metal flashings can be preserved with *Tinner's Red* or *Tinner's Green* paint or other suitable metal paint. The flashing should be painted with a coat of paint approximately every five years. The most common flashings that require regular painting include hips, ridges, valleys, chimney flashings, built-in gutter linings, pipe vent flashings, and any exposed ferrous-metal flashings.

3.3 SEALING OF FLASHING

A. Existing flashings shall be sealed with a high-grade caulk where needed, such as at mortar joints.

B. Loose flashings shall be riveted or else refastened in a manner that does not leave exposed or unsealed fasteners.

C. Missing flashing pieces shall be replaced with compatible flashing metal.

PART 4 — TEMPORARY EMERGENCY REPAIRS

As a last resort to stop water entry until proper repairs can be executed, temporary emergency repairs can be conducted on flashings when: a) painting no longer preserves the metal; b) the property owner can't afford to have the flashings replaced immediately; and c) entire roof replacement is neither desired nor needed.

4.1 GENERAL FLASHINGS

A. General flashing metal can be coated with a trowel-grade, or brush grade mastic or other sealant.

B. Reinforcing membrane, such as fiberglass mesh, is recommended to be included in all mastic repairs. When reinforcing membrane is used, a second layer of mastic must be applied over the membrane.

C. Avoid applying mastic or sealant over the exposed surface of the slates. Never apply a mastic repair over top of valley slates. Trowel mastic over the valley metal, but *underneath* the valley slates on either side of the valley.

4.2 BUILT-IN GUTTER LININGS

A. Built-in gutter linings can be temporarily preserved with a brush-grade roofing compound painted onto the cleaned gutter trough lining. Brush-grade or trowel-grade compounds should be layered with a reinforcing membrane, then a second layer of compound applied over top of the membrane. Roofing compounds should not be applied over top of any roofing slate surfaces.

B. Built-in gutter linings can also be temporarily preserved with a self-adhering membrane or other adhesive-applied membrane installed over the existing gutter lining, so long as the membrane is not applied over top of any slates.

C. Larger holes in the gutter lining should first be patched with trowel-grade roof cement and reinforcing membrane.

4.3 TEMPORARY SLATE REPAIR

A. Temporary repair of a missing or damaged slate can be accomplished by cutting a piece of non-ferrous sheet metal (copper, zinc, or dark aluminum), slightly wider than the damaged or missing slate, and sliding it in place *under* the defective slate or where the slate is missing. The flashing must slide underneath the overlying slates far enough to ensure correct headlap over the top of the flashing.

B. Broken or missing slates are not to be repaired with mastic or caulk applied on the exposed slate surface.

4.4 EXPOSED FASTENERS

Unsealed, ungasketed, exposed nail heads, screw heads, and other potentially leaking fasteners in existing flashings shall be sealed with a high-grade caulk, or the fasteners replaced with gasketed non-ferrous nails or screws, or with soldered fasteners, or with galvanically compatible rivets.

PART 5 — SLATE REPAIR

5.1 REPLACEMENT SLATES

A. All slate roof surfaces shall be inspected for cracked, broken, missing, perforated, face-nailed, screwed, caulked, or otherwise leaking slates.

B. Defective slates will be removed using appropriate tools such as a slate ripper. When using a slate ripper, the handle of the tool must be held down parallel to the roof surface when hammered on to pull slating nails. If the handle of the ripper is lifted off the roof during use, it can break the overlying slates.

C. Defective slates must be replaced with slates that match the existing slates in type, length, width, thickness, shape, color and, if necessary for a good match, age.

D. Refer to http://slateroofers.org/sources_salvaged_slate.html for a list of salvaged roofing slate suppliers.

5.2 NAIL AND BIB REPAIR METHOD

A. Replacement slates should be fastened with copper or stainless-steel smooth-shank roofing nails installed through the top of the slot between the two overlying slates, then covered with a minimum 16-ounce copper bib flashing, minimum 4 inches wide and minimum 6 inches long, slid underneath the overlying slates, but over the replacement nail head.

B. The bib flashing should overlap the nail head by three inches, and underlap the overlying slate by three inches.

C. The bib shall be bent slightly lengthwise to allow for a friction fit, and/or barbs may be cut into the sides of the bib to improve retention.

D. Alternatively, the same type of slating nails as on the existing roof can also be used to fasten the replacement slates. Many older slate roofs are installed with hot-dipped galvanized steel roofing nails, which can be used when installing replacement slates on these roofs.

E. Painted aluminum bib flashings (dark side facing out), minimum .019 gauge, are suitable for slate repair when the roof only has an estimated 50 or less years remaining.

5.3 SLATE HOOKS

Replacement slates may be installed using either copper or stainless-steel slate hooks.

5.4 STRAP HANGERS

Exposed sheet-metal strap hangers shall not be used as a replacement slate fastening mechanism.

5.5 CAULKING, MASTICS, AND ADHESIVES

A. Caulking, mastic, or adhesives shall not be used to repair the exposed face of the slates.

B. Caulking, mastic, or adhesives are not to be used to fasten slates without additional mechanical fasteners.

C. Caulking, mastic, or adhesives, in conjunction with mechanical fasteners, may be used to help fasten slates in limited circumstances, such as on windward rake edges, on ridges and hips that are subjected to chronic wind pressure, and on small pieces of slate that require additional anchoring.

D. Caulking, mastic, or adhesive is to be applied only underneath the slates, and only in conjunction with mechanical fastening. Otherwise, mastics and adhesives shall not be used underneath slates as they interfere with repair and maintenance.

5.6 FACE-NAILING

Replacement slates shall not be "face-nailed" or screwed. No exposed nail or screw heads, including gasketed or sealed heads, shall be visible on any repaired or replaced slates.

5.7 BIB FLASHING SLATE REPAIR

A. When a slate is split lengthwise (such as with a vertical hairline crack), or has a small hole in it, but is otherwise intact, it can be repaired with a non-corrosive sheet-metal bib flashing slid underneath the defective slate.

B. A slate with a vertical crack can be repaired by slipping an oversized bib flashing *underneath* the slate so that the bib extends from the bottom edge of the cracked slate up to its slating nails. A second bib flashing of the same size is also slid underneath the slot overlying the defective slate. The bib flashings should be a minimum of 6 inches in width and approximately as long as the length of the slate exposure plus three inches.

C. A slight lengthwise bend in the bib flashings, or else barbing the sides, will help to friction fit the bib into place.

5.8 SIDE LAP REPAIR

A. Where there is insufficient side lap, the deficiency can be repaired by sliding an oversized non-corrosive bib flashing underneath the side-butt joint that is too close to an underlying side-butt joint.

B. The bib flashing should be a minimum of 6 inches in width and approximately as long as the length of the slate exposure plus three inches.

C. A slight lengthwise bend in the bib flashing, or else barbing the sides, will help to friction fit the bib into place.

5.9 ROOF CANNIBALIZATION

A. When matching replacement slates cannot be obtained, identify a section of the existing roof that is:

1) Least visible from the ground,

2) Contains enough slates to repair the entire remaining roof,

3) Has slates that can be removed and used for the repairs.

B. Remove the slates from the identified area and use them for restoring the remaining roof.

C. The "cannibalized" roof section shall then be slated with either new or salvaged natural quarried roofing slates that match the original roof as closely as possible.

PART 6 — FLASHING REPAIR

6.1 REFERENCES

Refer to the Sheet Metal and Air Conditioning Contractors' National Association (SMACNA) *Architectural Sheet Metal Manual*, the Copper Development Association's (CDA) *Design Handbook*, and the Revere's *Copper and Common Sense* publication for additional flashing guidelines.

6.2 FLASHING INSPECTION

A. All flashings, including valleys, step and counter or cap flashings, parapet wall flashings, ridges and hip flashings, chimney flashings, dormer step flashings, apron flashings, finials, vent flashings, skylights and other roof penetrations, and gutters, are to be inspected for cracks, holes, splits, deterioration, exposed nail heads, and looseness.

B. Flashing metal that is perforated and leaking must be removed and replaced.

C. Special attention shall be paid to valley flashings, as these are more likely to be eroded over time than other flashings.

6.3 SOLDER JOINTS

A. Solder joints in flashings shall be inspected and resoldered or repaired as needed, when possible.

B. If solder joint failure is evident, the entire assembly should be examined for expansion and contraction design issues.

C. Refer to SMACNA, CDA and Revere publications for sheet metal expansion and contraction design recommendations.

6.4 MORTAR JOINTS

A. Mortar joints that have been cut out to allow for the insertion of new flashing may be sealed using a high-grade caulk/sealant.

B. Fully cut out mortar joints may be repointed with mortar.

C. Refer to National Parks Service (NPS) Preservation Brief *Repointing Mortar Joints in Historic Masonry Buildings.*

6.5 CHIMNEY FLASHING

A. Worn out or improperly installed chimney flashings shall be replaced in their entirety, including the front apron, the side step flashings, front, side and rear counter flashings, and back flashing and/or cricket flashings, per applicable building codes. Flashing material shall be minimum 16-ounce copper or 4-pound sheet lead.

B. All corner flashings on the chimney, where the chimney protrudes through the roof at the roofline, shall be installed to be water tight without sealant.

C. Chimneys that are no longer in use on old roofs can be removed to below the roofline when they are not a structural component of the roof.

1. It is imperative that existing chimneys be inspected before removal to ensure that

they are not in use for any reason (such as for a gas water heater), and are not structural.

2. The property owner should confirm, in writing, that the chimney is not structural and not in use for any reason, and must provide the roofer with written authorization to remove the chimney top, releasing him or her from liability.

3. Once the chimney top is removed, the resulting hole in the roof must be closed with matching roof sheathing and then slated to match the existing roof.

D. Refer to SMACNA, CDA and Revere publications for chimney flashing design recommendations.

6.6 POSITIVE FLASHING OVERLAP

All flashings are to be installed to have "positive overlap," such that higher flashings overlap on top of lower flashings, not underneath them. Refer to SMACNA, CDA and Revere publications for flashing overlap recommendations.

6.7 APRON FLASHINGS

Lower, exposed edges of apron flashings are to be fastened to the roof with cleats when fastening is required (no exposed fasteners are to be used).

6.8 REUSE OF EXISTING SLATES

Slates overlying flashings should be carefully removed and saved for reuse whenever possible.

6.9 VALLEY FLASHINGS

A. Open Valleys

1. Exposed valley metal that is pitted, split, leaking, tarred or otherwise compromised must be replaced in its entirety. The original, defective valley metal must be completely removed down to the bare roof deck. This is done by carefully removing several slates on either side of the valley to expose the existing flashing.

2. Slates overlying valley flashings must be carefully removed and saved for reuse whenever possible. Number the valley slate courses before removal. Numbering the slates makes the re-installation of the slates quicker and easier.

3. After removal of the original valley flashing, sweep the roof deck and check for nail heads or other obstacles that may impede the proper laying of the new flashing material. Re-nail the roof deck as needed. Install minimum 30 pound felt underlayment in the valley over the roof decking when required.

4. Valley metal should be a minimum of 16-ounce copper. The valley metal must be wide

enough to extend under the slates on either side of the valley by a minimum of 5 inches. Lower slope applications (e.g., 6:12 or less) may require increased overlapping.

5. Nails shall not penetrate more than one inch from the outer edges of the valley metal.

6. Alternatively, fasten the valley metal using copper cleats. Refer to SMACNA, CDA and Revere publications for cleated valley design recommendations.

7. Do not make valley sections longer than ten feet, otherwise expansion and contraction problems can occur.

8. Overlap valley sections six inches in most cases, but more overlap, as much as twelve inches, may be required in lower slope situations (e.g., 6:12 or less).

9. Instead of overlapping, valley sections may be seamed and soldered together, using cleats to attach them to the roof deck to allow for thermal movement.

10. Where two valleys meet at a top juncture, they can be folded, soldered, or lock-seamed. Refer to SMACNA, CDA and Revere publications for valley flashing design recommendations.

11. When two differing roof slopes come together in a valley (e.g., 6:12 on one side and 10:12 on the other), form the valley metal with a one-inch inverted "V" groove down the center.

12. When the valley is located between two steep symmetrical slopes, a simple straight-line bend down the center may be required.

13. Valley metal may be simply forced into the valley using a knee, creating a traditional rounded metal valley configuration. Once one side of the rounded metal valley is fastened to the roof deck, hold the metal down tight with your knee and fasten the other side, starting at one end and working toward the other. Traditional rounded valleys must be nailed rather than cleated.

14. Once new valley metal has been installed, reslate the valley. The numbered slates will be nailed back into the same place where they were removed.

15. If the original valley had a six-inch exposure (for example), mark lines three inches out from the center of the valley metal. When re-installing the slates, align them with the marks to create a straight valley appearance.

16. Valley exposure lines can be parallel, or they can be tapered so that they're wider at the bottom.

17. Any slates that were removed that are broken, tarred, cracked, perforated, or defective, should be replaced with sound, matching slates during the valley replacement process.

B. Closed Valleys

1. Closed valley flashings are to be replaced when any of the existing valley metal becomes pitted and leaking.

2. Slates overlying valley flashings should be carefully removed and saved for reuse whenever possible.

3. The valley slate courses are numbered before removal to facilitate the re-installation of the slates.

4. Closed valleys are flashed with a form of step flashing. A common method is to cut sheets of flashing material (minimum 16-ounce copper) and lay them in the valley with each slate course, on top of the slates, but covered by the overlying slates.

a. Closed valley flashing sections should overlap each other by a minimum of six inches, and extend out of the centerline of the valley a minimum of five inches.

b. The step flashing should extend above the top of the slates far enough to allow for nailing without penetrating the slates.

c. Slating nails shall not penetrate the flashings.

d. Flashing sizes vary according to slate size and roof slope.

e. The reinstalled valley slates shall be butted in the center.

f. For a closer fit in the center of the valley, cut the valley angle on the valley slates on the front face of the slate (rather than cutting them on the back, as is routine when cutting roofing slates).

5. Any slates that were removed that are broken, tarred, cracked, perforated, or defective, should be replaced with sound, matching slates during the valley replacement process.

6.10 PIPE FLASHINGS

A. Pipe flashings that are cracked, pitted, tarred, or otherwise defective, should be replaced in their entirety. The old pipe flashing is to be removed and a new flashing, either shop fabricated or purchased, is to be installed in its place. Non-corrosive metal pipe flashings are recommended.

B. The slates surrounding the pipe flashing must be carefully removed and saved for reuse, whenever possible.

C. The pipe flashing baseplate shall be sized to maintain a minimum of three inches of headlap with underlying and overlying slates, as well as minimum three inches sidelap with adjacent slates.

D. Any slates that were removed when replacing the pipe flashing that are broken, tarred, cracked, perforated, or defective, should be replaced with sound, matching slates during the flashing replacement process.

6.11 STEP AND COUNTER FLASHINGS

A. Step flashings may not be subjected to the exposure and wear that other flashings endure. For example, step flashings along parapet walls may still be serviceable although the valley flashings have worn out on the same roof. If so, the step flashings can remain original while other flashings are replaced.

B. Step flashings lower on a roof will wear out sooner than the upper ones and it may be necessary to only replace the lower flashings.

C. When step flashings are replaced, the procedure is the same as with most flashings. Slates overlying step flashings shall be carefully removed and saved for reuse. The original flashings are then removed in their entirety and replaced with minimum 16-ounce copper.

D. Step flashings shall be installed immediately underneath the overlying slates such that the bottom edge of the step flashing is flush with the bottom edge of the overlying slate.

E. Step flashings shall extend vertically underneath the second course of overlying slates a minimum of 3 inches. The lengths of the step flashings shall be equal to the length of the slate exposure plus a minimum 3 inches, thereby allowing the step flashings to overlap each other by a minimum of 3 inches (4 inches when the roof slope is less than 8:12, two inches when the slope is 20:12 or greater).

F. Step flashings shall extend a minimum of 4 inches onto the roof and a minimum of 4 inches vertically, wherever possible.

G. Nail the step flashing to the roof above the underlying slate so that the slating nails do not penetrate the step flashing.

H. If replacement of the overlying counter flashing or cap flashing is necessary, remove the original counter flashings by pulling or prying them out of the mortar joints or anchor points.

1. Replace counter flashings as needed using minimum 16-ounce copper or 4-pound sheet lead.
2. Wedge the counter flashings into mortar joints at least 3/4 inches deep, using galvanically compatible metal wedges, such as lead or copper.
3. Mortar joints that have been cut out to allow for the insertion of the new flashing may be resealed using a high-grade caulk/sealant.
4. Fully cut out joints can be repointed with mortar.
5. Refer to National Parks Service (NPS) Preservation Brief *Repointing Mortar Joints in Historic Masonry Buildings.*

6.12 RIDGES AND HIPS

A. Slate ridges/hips are repaired by removing and replacing the defective slate(s). These can sometimes be pulled out using a slate ripper. Replacement slates may be installed using a pair of slate hooks rather than nails. This is a situation where an adhesive underneath the replacement slate may be beneficial.

B. Saddle slate ridges may need to be completely or partially disassembled and reassembled to avoid face-nailing.

C. Worn metal ridges and hips are to be completely removed and replaced in their entirety. There are often no slates overlying metal ridges and hips, but the underlying slates are typically in need of repair. These repairs should be conducted when the metal is removed.

D. When replacing metal ridges and hips, it is recommended to use wider replacement metal to cover the old nail holes, if the original ridges and hips had been nailed.

E. It is recommended that new ridge/hip metal be installed with concealed fasteners such as cleats, not with exposed nails.

F. If ridge/hip metal is being nailed into place, the nail heads should be gasketed.

G. New ridge/hip metal must be sized to maintain correct headlap with underlying slate.

6.13 GENERAL ROOF PENETRATIONS

There may be other roof penetrations that have not already been discussed here that may need flashing replacement. These can include such things as air vents, skylights, and cupolas. Generally, these can be reflashed in the same manner as chimneys and the other flashing systems discussed herein.

6.14 THROUGH-WALL FLASHINGS

If non-ferrous metal through-wall flashings are found properly installed and in good condition, they may be reused when underlying flashings are replaced.

6.15 RECYCLING FLASHINGS

Old flashing metal, when removed for replacement, should be taken to a scrap yard or otherwise recycled, whenever possible.

PART 7 — GUTTERS

7.1 BUILT-IN GUTTERS

A. Built-in gutters include box gutters, Yankee gutters, Philadelphia gutters, pole gutters, flush gutters and standing gutters, among others. Built-in gutters are built into and below the roof plane, or on the plane of the roof, usually at or near the eaves. These types of gutters are common on older slate roofs.

B. When built-in gutters wear out and leak, the metal linings must be removed and replaced in their entirety.

C. Built-in gutter liner replacement can be delayed with temporary or emergency patching as discussed in the Maintenance and Temporary Emergency Repairs sections of these guidelines.

7.2 HANGING GUTTERS

A. Hanging gutters, including half-round, ogee, and k-style gutters, are installed on the exterior of the structure at the eaves. When they wear out and leak, are damaged, or installed incorrectly, they may need to be removed and replaced in their entirety.

B. Gutter hangers shall not be face-nailed or screwed through the roofing slate, but must be installed directly onto the roof deck underneath the slates, or to the fascia.

C. Exposed gutter hangers face-nailed through slates must be removed, correct hangers properly installed underneath the slates, and any damaged slates repaired.

D. It is recommended that the metal type of hanging gutters be galvanically compatible with roof valleys and other roof flashings.

PART 8 — SNOW RETENTION SYSTEMS

8.1 SNOW RETENTION SYSTEM LAYOUTS

A. Snow guard layouts should be designed by a qualified design professional.

B. Snow guard quantities and layout should be based on conditions specific to the geographic area, roof slope, rafter length, etc.

C. Refer to manufacturer's recommendations for snow guard installations.

8.2 FAILURE AND REPLACEMENT

A. Snow retention systems may fail with age, or when incorrectly designed, or when the snow guards are installed in insufficient quantities.

B. Inadequate snow retention systems can be removed and replaced in their entirety, or supplemented with additional snow guards or snow rails.

Slate Grades

ASTM S1: Expected service life in excess of 75 years.
ASTM S2: Expected service life 40-75 years.
ASTM S3: Expected service life 20-40 years.

Thicknesses

Standards; 3/16 inch (5 mm) to 1/4 inch (7 mm)
Quarters; 1/4 inch (7 mm) to 3/8 inch (10 mm)
Heavies; 3/8 inch (10 mm) to 1/2 inch (13 mm)
Extra Heavies; 1/2 inch (13 mm) to 3/4 inch (19 mm)

Colors (North American)

Unfading Black
Semi-Weathering Gray/Black
Unfading Gray
Semi-Weathering Gray
Unfading Purple
Semi-Weathering Purple
Unfading Mottled Green and Purple
Unfading Green
Semi-Weathering Green
Unfading Red

NUMBER OF SLATES AND NAILS

FOR 100 SQUARE FEET OF ROOF

FOR STANDARD THICKNESS (3/16" TO 1/4") SLATES USING 1.5" NAILS

SIZE OF SLATE (INCHES)	EXPOSED LENGTH (INCHES)	# SLATES 3" HEADLAP	APPROXIMATE WEIGHT OF NAILS (POUNDS) GALV.	COPPER 11 GA.	STAINLESS 11 GA	SPACING OF LATH
6x10	3 1/2	686	7 7/8	8.85	8.07	3 1/2
7x10	3 1/2	588	6 3/4	7.58	6.91	3 1/2
8x10	3 1/2	514	5 7/8	6.63	6.04	3 1/2
6x12	4 1/2	533	6	6.87	6.27	4 1/2
7x12	4 1/2	457	5 1/4	5.89	5.37	4 1/2
8x12	4 1/2	400	4 5/8	5.16	4.70	4 1/2
9x12	4 1/2	355	4 1/8	4.58	4.18	4 1/2
10x12	4 1/2	320	3 5/8	4.13	3.76	4 1/2
7x14	5 1/2	374	4 1/4	4.82	4.40	5 1/2
8x14	5 1/2	327	3 3/4	4.22	3.85	5 1/2
9x14	5 1/2	290	3 3/8	3.74	3.41	5 1/2
10x14	5 1/2	261	3	3.37	3.07	5 1/2
12x14	5 1/2	218	2 1/2	2.81	2.56	5 1/2
8x16	6 1/2	277	3 1/8	3.57	3.26	6 1/2
9x16	6 1/2	246	3	3.17	2.89	6 1/2
10x16	6 1/2	222	2 1/2	2.86	2.61	6 1/2
12x16	6 1/2	185	2 1/8	2.39	2.18	6 1/2
9x18	7 1/2	213	2 1/2	2.75	2.50	7 1/2
10x18	7 1/2	192	2 1/4	2.48	2.26	7 1/2
11x18	7 1/2	175	2	2.26	2.06	7 1/2
12x18	7 1/2	160	1 7/8	2.06	1.88	7 1/2
10x20	8 1/2	170	2 3/8	2.19	2.00	8 1/2
11x20	8 1/2	154	2 1/8	1.99	1.81	8 1/2
12x20	8 1/2	141	2	1.82	1.66	8 1/2
14x20	8 1/2	121	1 7/8	1.56	1.42	8 1/2
11x22	9 1/2	138	2	1.78	1.62	9 1/2
12x22	9 1/2	126	1 3/4	1.63	1.48	9 1/2
14x22	9 1/2	109	1 1/2	1.41	1.28	9 1/2
12x24	10 1/2	114	1 5/8	1.47	1.34	10 1/2
14x24	10 1/2	98	1 3/8	1.26	1.15	10 1/2

[Sources: *Radford's Estimating and Contracting* (1913), p. 252; and Slate Roofs (1926), p. 12; # slates for 4" headlap from *Slating and Tiling* by J. Millar, 1937, pp. 21-22]

Nails

Correct nail length is twice the thickness of the slates plus one inch (nail length=slate thicknessX2+1").

Recommended nails for new slate installations include solid copper or stainless steel. Type 304 stainless nails are made of nickel chromium stainless steel. Type 316 contains slightly more nickel than Type 304 as well as 2-3% molybdenum, making it more resistant to corrosion caused by seawater or sulphuric acid. Both types of stainless show negligible weight loss and no corrosion when tested under one year of accelerated exposure. Under the same exposure, copper nails show a 26-28% weight loss and visible corrosion, while aluminum nails show a 32% weight loss and pronounced corrosion, and electrogalvanized nails show a 100% weight loss and complete corrosion.

(Source: Swan Secure Products, Fasten Data 105)

LENGTH (INCHES)	GAUGE	SHANK DIA. (INCHES)	COPPER HEAD DIA.	COPPER COUNT/LB	STAINLESS HEAD DIA.	STAINLESS COUNT/LB
3/4	12	.109	21/64	360	5/16	380
1	12	.109	21/64	288	5/16	348
1.25	12	.109	21/64	236	5/16	255
1.5	12	.109	21/64	202	5/16	218
1.75	12	.109	21/64	176	5/16	189
2	12	.109	21/64	156	5/16	170
3/4	11	.120	3/8	302	11/32	323
1	11	.120	3/8	229	11/32	244
1.25	11	.120	3/8	187	11/32	199
1.5	11	.120	3/8	155	11/32	170
1.75	11	.120	3/8	139	11/32	149
2	11	.120	3/8	124	11/32	136
1	10	.134	7/16	187	3/8	212
1.25	10	.134	7/16	147	3/8	166
1.5	10	.134	7/16	123	3/8	139
1.75	10	.134	7/16	112	3/8	126
2	10	.134	7/16	93	3/8	105
2.5	10	.134	7/16	81	3/8	91
3	10	.134	7/16	67	3/8	78

ASTM References

ASTM A167 — Type 304 Terne coated stainless steel, 0.015 inch (0.38 mm) thick stainless steel core material, coated with 0.092 lb/sq ft (450 g/sq m) Terne alloy on both sides.
ASTM A666 — Type 304 stainless steel, soft annealed 2D finish (unless harder temper is required for forming or performance), 0.0156 inch (0.4 mm)
ASTM B101 — Specification for Lead-Coated Copper Sheet and Strip for Building Construction
ASTM B370 — Specification for Copper Sheet and Strip for Construction
ASTM B749 — Lead and Lead Alloy Strip, Sheet, and Plate Products
ASTM C406 — Specification for Roofing Slate
ASTM C920 — Specification for Elastomeric Joint Sealants
ASTM C1311 — Specification for Solvent Release Sealants
ASTM D312 — Specification for Asphalt Used in Roofing
ASTM D1079 — Standard Terminology Relating to Roofing, Waterproofing, and Bituminous Materials
ASTM D1970 — Specification for Self-Adhering Polymer Modified Bituminous Sheet Materials Used as Steep Roofing Underlayment for Ice Dam Protection
ASTM D2626 — Specification for Asphalt-Saturated and Coated Organic Felt Base Roofing Sheet
ASTM D2822 — Specification for Asphalt Roof Cement, Asbestos-Containing
ASTM D3019 — Specification for Lap Cement Used with Asphalt Roll Roofing, Non Fibered, Asbestos, Fibered, and Non Asbestos Fibered
ASTM D4022 — Coal Tar Roof Cement, Asbestos Containing
ASTM D4586 — Specification for Asphalt Roof Cement, Asbestos-Free.
ASTM D4869 — Specification for Asphalt-Saturated Organic Felt Underlayment for Steep Slope Roofing.
ASTM F1667 — Standard for Driven Fasteners

Slate Exposures and Headlaps

1. Slate Size (in.)
2. Exposure (3" Headlap)
3. Slates/Square (3" Headlap)
4. Exposure (4" Headlap)
5. Slates/Square (4" Headlap)

1	2 INCHES	3	4 INCHES	5
6x10	3 1/2	686	3	800
7x10	3 1/2	588	3	686
8x10	3 1/2	514	3	600
6x12	4 1/2	533	4	600
7x12	4 1/2	457	4	515
8x12	4 1/2	400	4	450
9x12	4 1/2	355	4	400
10x12	4 1/2	320	4	360
7x14	5 1/2	374	5	412
8x14	5 1/2	327	5	360
9x14	5 1/2	290	5	320
10x14	5 1/2	261	5	288
12x14	5 1/2	218	5	240
8x16	6 1/2	277	6	300
9x16	6 1/2	246	6	256
10x16	6 1/2	222	6	230
12x16	6 1/2	185	6	192
9x18	7 1/2	213	7	221
10x18	7 1/2	192	7	199
11x18	7 1/2	175	7	187
12x18	7 1/2	160	7	171
10x20	8 1/2	170	8	180
11x20	8 1/2	154	8	164
12x20	8 1/2	141	8	150
14x20	8 1/2	121	8	129
11x22	9 1/2	138	9	146
12x22	9 1/2	126	9	134
14x22	9 1/2	109	9	115
12x24	10 1/2	114	10	120
14x24	10 1/2	98	10	103

Source: Slate Roofing Contractors Association of North America, Inc. (SlateRoofers.org), "General Installation Guidelines for Natural Quarried Roofing Slate" (2012).

Roofing Slate Sizes Inches and Metric

INCHES	INCHES	
24 L X	14 W	(610 mm X 356mm)
24 L X	12 W	(610 mm X 305 mm)
22 L X	12 W	(559 mm X 305 mm)
22 L X	11 W	(559 mm X 279 mm)
20 L X	14 W	(508 mm X 356 mm)
20 L X	12 W	(508 mm X 305 mm)
20 L X	11 W	(508 mm X 279 mm)
20 L X	10 W	(508 mm X 254 mm)
18 L X	14 W	(457 mm X 356 mm)
18 L X	12 W	(457 mm X 305 mm)
18 L X	11 W	(457 mm X 279 mm)
18 L X	10 W	(457 mm X 254 mm)
18 L X	9 W	(457 mm X 229 mm)
16 L X	14 W	(406 mm X 356 mm)
16 L X	12 W	(406 mm X 305 mm)
16 L X	11 W	(406 mm X 279 mm)
16 L X	10 W	(406 mm X 254 mm)
16 L X	9 W	(406 mm X 229 mm)
16 L X	8 W	(406 mm X 203 mm)
14 L X	10 W	(356 mm X 254 mm)
14 L X	9 W	(356 mm X 229 mm)
14 L X	8 W	(356 mm X 203 mm)
14 L X	7 W	(356 mm X 178 mm)
12 L X	10 W	(305 mm X 254 mm)
12 L X	9 W	(305 mm X 229 mm)
12 L X	8 W	(305 mm X 203 mm)
12 L X	7 W	(305 mm X 178 mm)
12 L X	6 W	(305 mm X 152 mm)

THICKNESSES

3/16" = 4.76 mm	5/8" = 15.87 mm
1/4" = 6.35 mm	3/4" = 19.05 mm
3/8" = 9.525 mm	7/8" = 22.225 mm
1/2" = 12.7 mm	1" = 25.4 mm

Source: Slate Roofing Contractors Association of North America, Inc. (SlateRoofers.org), "General Installation Guidelines for Natural Quarried Roofing Slate" (2012).

TOP 10 MISTAKES MADE WHEN INSTALLING NEW SLATE ROOFS

1) Lack of information: You have this book in your hands, so you're way ahead of the game. Check out SlateRoofCentral.com, SlateRoofWarehouse.com, TraditionalRoofing.com and SlateRoofers.org.

2) Know your slate! There are many different types of slate with differing characteristics and longevities. Do the research.

3) Have detailed contract documents! Every detail about the slate roof installation should be included in the contract documents — type, size and origin of the slate; type, length and gauge of the nails; type and installation style of underlayment; type and size of cant strip; headlap; flashing specifications; number of squares to be installed; slate installation style, and many other details.

4) Lack of headlap: Lack of adequate headlap spells disaster for a slate roof.

5) Bad flashing work: There are two things that keep water from penetrating a slate roof: the slates and the flashings. Not only must the flashing metal be of adequate type and gauge, but it must be installed correctly.

6) No roof consultant was used on the job: Don't spend a half million dollars on a new roof without hiring a consultant to at least review your contracts.

7) Contractors walk on the slate: The slate roof must be properly staged so the roofers are working off roof ladders and roof scaffolds. Good slaters won't walk on a slate roof unless they have to.

8) Poor roof deck: If you're going to install slate on half-inch plywood, you don't have a clue.

9) Emphasis on underlayment: The main purpose of the underlayment is to keep the water out of the building until the slate and flashings are installed. The roof will vastly outlast the underlayment and will still be watertight for generations afterward.

10) Inexperienced roofing contractors: Some contractors cannot be trusted to give sound and honest advice or information. Screen contractors for slate roofing purposes by utilizing professional trade associations such as the Slate Roofing Contractors Association of North America, Inc. (SlateRoofers.org).

WHEN STARTING TO SLATE A ROOF 10 QUICK-REFERENCE STEPS

1) Make sure that the fascia is completely installed before slating the roof and that the ends of the sheathing boards are firmly nailed.

2) Felt over the solid board decking (avoid plywood) with minimum 30 lb. roofing felt, lapped at least three inches at the top. Felt paper is optional, but recommended.

3) Nail or screw a wooden cant strip at the bottom edge of the lowest sheathing board. It should be 3/8" to 1/2" thick (for standard thickness slates) and at least an inch wide (eight foot lengths are convenient). Cedar or redwood is ideal (cedar shim shingles will also work), but the same glue-free lumber as the roof decking will do fine. The cant strip will be thicker when installing thicker slates.

4) Chalk a horizontal line on the felt paper for the top edge of the starter slates, deducting 1½ inches for the slate overhang. Then chalk a line for the first full row of slates, now measuring up the roof the length of the slate minus 1½ inches for the overhang. Lay starter slates back side out. Make sure you have a minimum of 3 inches headlap on the starter course, four inches when the slope is below 8:12.

5) Measure up the remainder of the roof equal distances equivalent to the *exposure* of the slate, and chalk lines accordingly. Make sure your second full row of slates will overlap the starter row by a minimum of three or four inches. If not, drop the top of the second row down to where you need it to be, *then* chalk the rest of the roof with the exposure measurement. If slating in an ice-dam-prone area, lay the bottom three feet of the roof with a 5" headlap.

Exposure is determined by subtracting the headlap from the total length of the slate, then dividing the remainder in half. For example, a 20" slate with a 3" headlap will have an 8.5" exposure (20 minus 3 = 17, divided in half = 8.5"). Although 3" headlap is standard, 4" headlap is required for lower-slope slate roofs and maybe more for ice-dam-prone sections.

6) Do not bed any slates in roof cement or caulk. Adhesives make it very difficult to repair the roof in the future. Instead, a pair of smooth-shank copper or stainless steel nails per slate will ensure the secure attachment of all slates to the roof. Don't nail the slates *too* tightly; let them hang snugly on the roof. Do make sure the nail heads are set into the slate, because nails that stick up can wear a hole in the overlying slate and cause a leak.

7) Tap a couple of temporary nails into the side of the fascia on the gable end, one at the top and one at the bottom, and run a string up the edge of the roof positioned one inch out from the fascia. Use the string as a guide to align the outside edge of the slates as you nail them into place. Remove the string when you're done. Alternatively, chalk a vertical line on the felt up the edge of the roof as a guide for the end slates.

8) The butted sides of the slates on the first row must be staggered at least 3" laterally from the butted sides of the starter slates. All slates should have a minimum 3" lateral overlap in relation to the row of slates above or below.

9) You can work the first half dozen rows from a ground ladder or ground scaffold, then nail roof jacks and planks along the bottom of the roof and work up from there. Use more jacks and planks as needed. Leave a slate out periodically in order to have a place to nail the roof jack. The missing slates can be installed later with a stainless steel slate hook.

10) Remember that all rules can have exceptions.

Avoid These
21 Common Errors

1) SIZE OF SLATE: The smaller the slate shingle, the more that are required to cover one hundred square feet of roof, a "square." The largest standard size, 14"x24", requires 98 shingles per square. The smallest standard size, 6"x12", requires 533 per square. Since each shingle is attached to the roof deck with two nails, the smaller slate will require much more labor during installation. Do not purchase small slates just because they're cheaper because your labor costs will go way up.

2) TYPE OF SLATE: Buy a slate that is tried and proven, with many decades of outstanding performance in the field, manufactured by a company that takes pride in its product. Most American and Canadian slates fall into that category. There are some foreign slates with little or no history of performance. You don't want to have to remove and reslate an entire roof at great expense because you bought bad slates. Again, cost should not be the deciding factor when selecting roofing slates.

3) OPENING PALLETS: Even if the slate is all from the same source, different pallets can contain different shades of slates due to the location of the rock strata or other factors related to the quarrying of the material. Therefore, when the slate is delivered to the site, all of the pallets should be opened from the outset and slates taken simultaneously from each pallet in order to blend the entire inventory and create a pleasant appearance on the roof. It is a mistake to start with one pallet, install those slates, then open another pallet, install those slates, etc.

4) CULL OUT BAD SLATES: Poor quality slates can originate from any quarry. Slates with pyrite inclusions (Figure 1), curvature, thick tops, nail holes at the wrong end, bumps, knots and knuckles, should be set aside and not installed on the roof.

5) BUY FROM A CONSCIENTIOUS SOURCE: There are other nuances related to the manufacture of roofing slates that can affect quality and longevity. Direction of grain and nail hole placement are two examples. A good manufacturer will be aware of these details and strive to produce top-quality slate shingles. It pays to buy from such a source.

6) DON'T REINVENT THE WHEEL: Styles and methods of slate installation can vary greatly, but one important point needs to be emphasized: if it's *longevity* that's desired, then the installation system should be based upon known methods and materials that have proven themselves. A solid, glue-free wood roof deck, with the emphasis on correct slate and flashing installation and not on underlayment, has proven to provide exceptional longevity.

7) GET THE STARTER COURSE RIGHT: The starter course is the very first row of slates to be installed. Starter slates are invisible once the roof is completed because they're hidden underneath the first course of slate. The starter course slate should be installed face down, unlike all the other slates on the roof, which are installed faceup. Read more detail about starter course installation later in this chapter.

8) GET THE HEADLAP RIGHT: If the headlap is missing or inadequate, the roof can be condemned (Figure 2). Headlap is the overlap on each course of slate by the second course above it. This overlap is what prevents the roof from leaking. Three inches is standard, but headlap can vary according to the slope of the roof. Read more detail about this later in this chapter.

9) WATCH YOUR LATERAL OVERLAPS: Lateral overlaps, or sidelaps, should also be 3" minimum. Incorrectly placed sidelaps can leave the butt-joint directly over the slating nails, and create a leaking roof. Figure 3 shows an incorrectly installed roof.

10) AVOID OVERNAILING AND UNDERNAILING: Slate nail holes are meant to be crater-shaped to allow the nail head to sit down inside the slate. Otherwise, the nail head will protrude above the slate and rub against the overlying slate, eventually creating a hole (Figure 4). When the nail isn't driven far enough, this is known

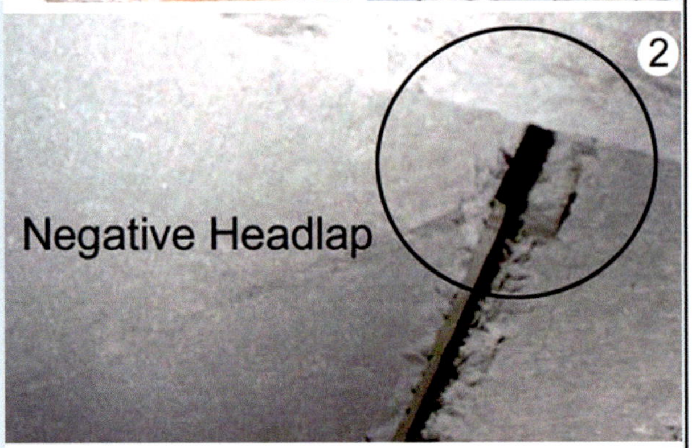

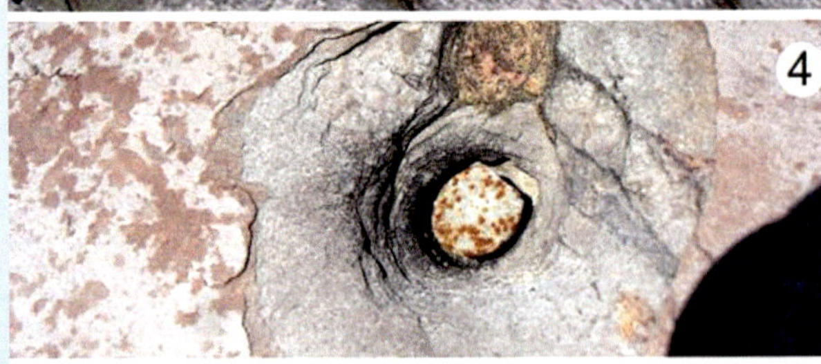

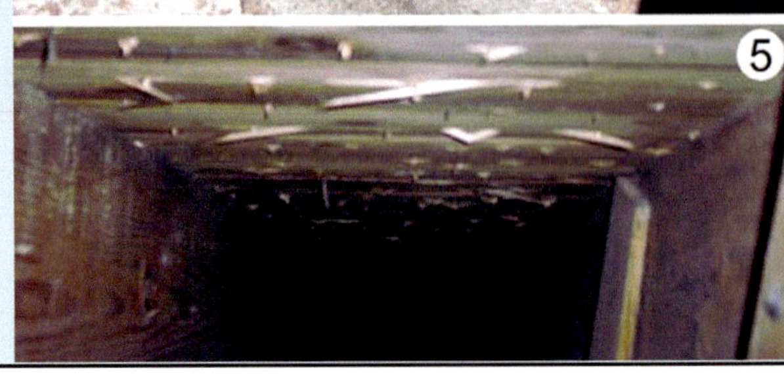

as undernailing. Overnailing is when the nail is driven too far and breaks through the slate. This leaves the slate hanging on one nail or sliding out altogether (Figure 6).

11) USE THE RIGHT NAIL LENGTH: The slating nails should barely penetrate through the bottom of the roof deck boards, not break through the back of the boards and splinter out the wood (Figure 5). This reduces the board thickness where the nail is located and undermines the effective holding power of the nail.

12) STAGE THE ROOF CORRECTLY: A common error is roofers walking on the slates during installation. An experienced slating crew will make every effort to keep off the slates during installation by staging the roof correctly. Slates damaged by foot traffic may not fall apart immediately, but hundreds can fall off a new slate roof within five years after installation (Figure 7).

13) INSTALL FLASHING CORRECTLY: Flashings provide all sorts of opportunities for error. Uphill flashing should always lap on top of downhill flashing. When lower flashings lap on top of higher flashings, water can enter the roof. Corner flashings on roof penetrations must be correctly folded, seamed, or else soldered to prevent leakage.

14) ALLOW FOR METAL EXPANSION: Built-in gutter systems require expansion joints, otherwise, the solder joints can fail. Flat-lock copper roofs also require expansion joints when over 30 feet. Refer to SMACNA for details.

15) DON'T USE MAGNETIC RIVETS: Many roofing supply outlets sell copper rivets with copper-plated steel mandrels. When a rivet is installed, the steel mandrel breaks off inside the rivet, rusts, and creates a hole in the flashing (Figure 8). Copper rivets should have brass mandrels, not steel. If they're magnetic, throw them out.

16) USE COMPATIBLE METALS: A common sight on new roof installations is metal incompatibility, often steel and copper used together. When exposed to moisture, the copper will cause steel fasteners to degrade at an accelerated rate (Figure 9).

17) DON'T SOLDER WITH AN OPEN FLAME: Open-flame torches should never be used to solder flashings, so if you see a worker on a slate roof soldering a box gutter, for example, with an open flame plumber's torch, *say something*. The flame is too hot and will ignite the felt paper or rosin sheet, which will then smolder underneath the metal. There is no way to get to the fire without first ripping out the flashing. The correct tool for a soldering job is a closed-flame soldering device, or a heavy-duty electric soldering iron. Open flame devices can be used only on *external* hanging gutters.

18) HANG GUTTERS BELOW THE PLANE OF THE ROOF: The outer edge of external gutters should be below the plane of the roof in locations where snow or ice could slide down the roof and knock them off. See Figure 10.

19) USE ENOUGH SNOW GUARDS: Snow retention systems can be a source of problems when they're installed incorrectly. Make sure you use enough of them. Otherwise, they won't be able to hold the weight of the ice and snow and will rip out, taking slates with them. Follow the manufacturer's guidelines when installing them.

20) USE THE CORRECT TOOLS: Slates should be cut with slate cutters. A slate cutter will leave a beveled edge on the shingle, allowing it to match all the other shingles on the roof, all of which have beveled edges. Slate hammers all have the necessary punch for punching nail holes in roof slates. Slate rippers are also an essential tool. Make sure you use them.

21) INSTALL *UNDER*LAYMENT *UNDER* SLATES: Do not install underlayment over top of each course of slates before the next course is laid, and certainly never use self-adhesive underlayment on top of slate courses. There is no way to repair a slate roof that has been glued together, and even felt on top of slates makes repair a nightmare. The beauty of traditional slate roofs is that they are like cars; you can remove any part and replace it, and in so doing, you can keep such a roof alive for centuries.

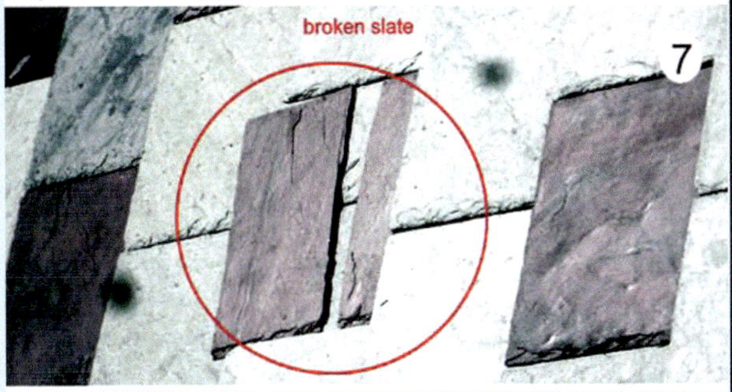

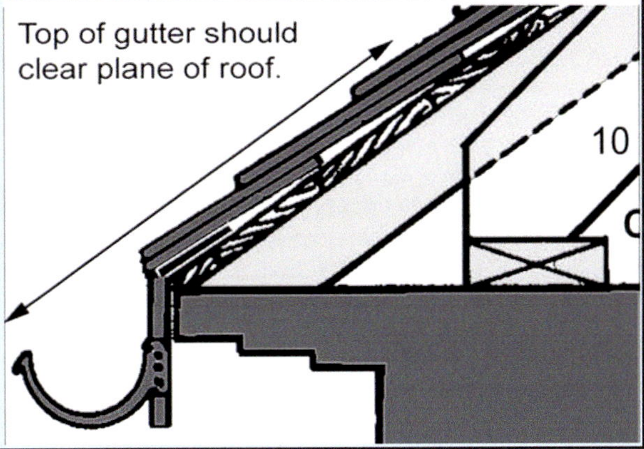

Five Common Mistakes Made When Installing

Slate Starter Courses

1) THE STARTER COURSE IS FACEUP. If the starter course is laid with the face side up, the installer is inexperienced, and it's likely there are additional mistakes in the roof installation. The starter slates are the only ones on the roof installed with the face down and the back up (Figure 1). All other slates on the roof are installed with the face up to expose the beveled edges of the slate (the face of the slate is the side that shows the beveled edge — the back doesn't). The reason the starter course is laid face down is because the edge of the starter slate merges cleanly with the edge of the first course of slate (Figure 2). This just looks better, but it doesn't have any real effect on the long-term functionality of the roof.

Furthermore, when ordering roofing slates from the quarry, order the starter slates separate from the field slates (rather than use the field slates as starter slates). A good slate supplier will make sure the starter slates are punched for nail holes only along the top of the slate on the front (rather than on the back as is done on the field slates). This allows the slate to be laid face down.

2) THE CANT IS MISSING. The starter course must be angled on the roof to match the angle of the slates in the field of the roof. No slate lies flat on the roof because every slate is overlapping other slates (Figure 3). Since the starter course doesn't have a course of slate underneath it, something must be installed to create the proper angle. Traditionally, this has been done by nailing a cant or shim strip made from wood under the bottom edge of the starter course (Figure 4). Alternatively, the fascia board can be raised slightly to raise the bottom of the starter course. Also, if a metal drip edge is used, the drip edge can be formed with an inverted "v" groove that functions in the same manner as a cant or shim. Other types of cants have been employed over the years, such as cedar shakes, plaster lath, etc. If the shim is left out, the slates may not lie correctly on the roof. This is usually manifested as a gap underneath the bottom courses of slate.

3) SIDELAPS ARE INADEQUATE. The sidelaps, or lateral overlaps between the side-butts of the starter slates and the first course of slates, should be a minimum of 3". On a poorly installed starter course, these lateral overlaps will measure an inch or less (Figure 5). When the side-butts of the first course are too close to the side-butts of the starter course, a path is created for water entry into the building. Where there is no sidelap, there is a certain leak.

4) LACK OF HEADLAP. This is a common mistake made when the field slates are turned sideways and used for starters and no adjustment is made on the first course exposure to allow for adequate headlap on the starter (Figure 6). The height (distance up the roof) of the starter slate must be at least 3" greater than the vertical exposure on the first course of slate. The exposure on the first course of slates may be reduced by dropping the chalk line for the top of the second course of slates to increase the headlap on the starter course. This little detail is often overlooked by inexperienced roofers. Since the bottom of the roof has more water running over it than virtually any other part, it is

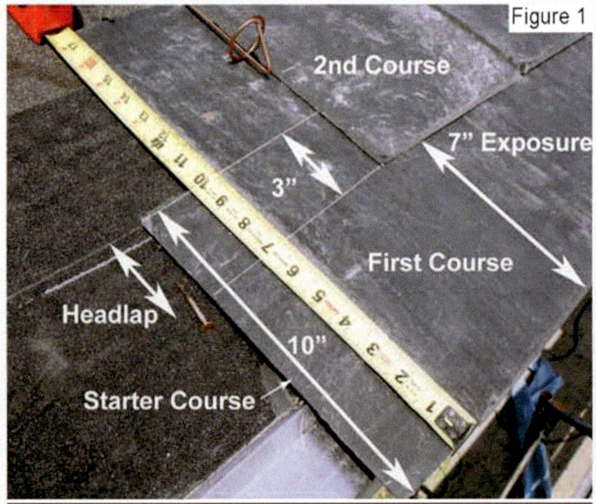

The starter course is supposed to be laid with the back side of the slate (the side without the beveled edge) facing upward, as shown above, and below right. When installed correctly, the drip edge bevels merge to a point, as shown below in Figure 2, right side. When installed incorrectly, as shown below left, it's an indication of an inexperienced installer and does not bode well for the rest of the installation.

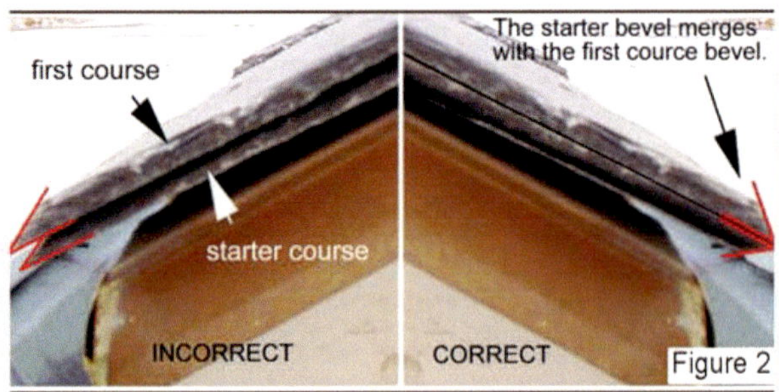

No slate lies flat on the roof. They are all angled, so the starter slate must have a shim or cant underneath it to create the proper angle.

An example of a traditional wooden cant strip on a slate roof being installed over a standing seam copper roof.

imperative that the headlap on the bottom courses of slate be correct. If anything, the slates on the bottom of the roof should have more headlap than the field slates, not less.

5) THE STARTER COURSE IS INSTALLED AHEAD OF THE FIRST COURSE. It is important that the starter course and the first course of slates be installed at the same time, although many installers tend to run the starter course out ahead of the first course (Figure 8). The exception to this rule is when the starter slates are the same width as the field slates (or a factor thereof), and the starter slates are only punched for nail holes on the top edge; then the starter course can be laid out ahead of the first course. In all other circumstances, such as when installing random width slates or using field slates turned sideways as starters, the starter course and the first course should be installed concurrently. This is the only way you can guarantee correct sidelaps and make sure you don't have stray nail holes in the starter slates in the joint between the first course slates, creating a leak (Figure 7). When you lay the starter and first courses concurrently, you can correctly trim each starter slate to avoid these problems.

For example, let's say your field slates are 10"x20". For a 3" headlap, you will need an 8.5" exposure. However, if you turn the slates sideways and use them for starter slates, you will only have 10" of slate going up the roof on the starter course. An 8.5" exposure on the first course of slates, then, would only leave a headlap of 1.5" on the starter course — a mistake. The exposure on the first course of slate must instead be dropped to 7" to create the 3" headlap on the starter course.

Alternatively, you could use a 12" high starter slate, thereby guaranteeing a 3.5" headlap on the starter course and an 8.5" exposure on all field slates. In fact, when you ordered the 10x20 field slates from the quarry, you should have also ordered 10" wide and 12" long (high) pre-punched starter slates. That would ensure you an easy, foolproof job, and you could go ahead and lay the starters out ahead of the first course.

In addition, if you use the field slates turned sideways as starter slates, then you have to punch an extra nail hole on one side of each starter slate to nail the slate to the roof sideways since all slates on the roof must be nailed with two nails. This will leave you with a "stray hole" (an original nail hole) along the bottom of each slate, which can create leaking problems if it falls in the slot between two overlying first course slates. So even though the 20" wide by 10" high starter slates are a factor of the width of the 10" wide field slates, they cannot be laid out in advance of the field slates without risking a stray hole falling in an overlying joint. The starters and the first course must be laid concurrently in this situation.

The starter course, even though invisible, is arguably one of the most important courses on a slate roof. If it's done right, you're off to a good start. If it's done wrong, you've created a slate roof with built-in long-term headaches. Some new installations even go into litigation because of problems with the starter course. If you do a little homework, educate yourself, and see and understand these potential problems before you start installing a slate roof, you will be doing both yourself and your roofing clients a big favor in the long run.

Figure 5

Figure 6

Figure 7

Note proximity of nail head to side-butt joint of first course. The nail is too close to the joint. This could be avoided by installing the starter course and first course at the same time. Although this nail will be covered by the 2nd course headlap, it still creates a weak point that could be subject to water penetration during, for example, ice-dam conditions.

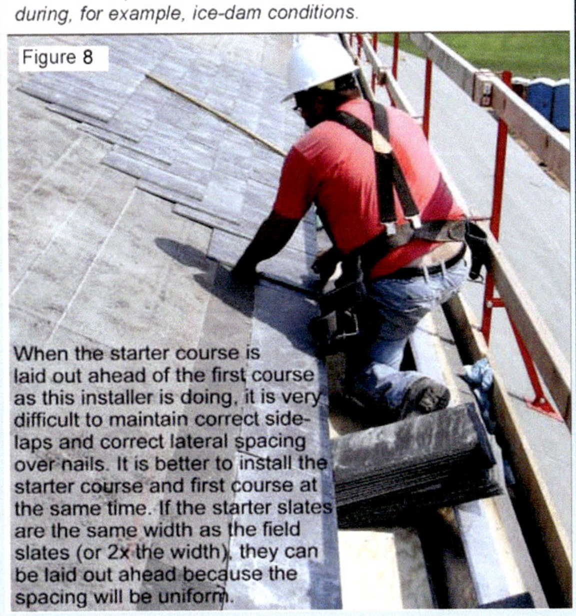

Figure 8

Headlap

Insufficient headlap. Zero headlap. Negative headlap. These are all words that should send a chill up the spine of any good slater. There are few details on a slate roof installation that, if done wrong, will cause the entire roof to be condemned and have to be removed and done all over again. Headlap is one of them.

As a slate roof consultant, the author has had the opportunity to travel the country and look at newly installed slate roofs, many unfortunately riddled with glaring installation mistakes. If the headlap is OK, then the roof installation has hope. If the headlap is missing or inadequate, the roof is doomed.

What exactly is headlap? Many contractors will take on a slate roof installation project with little or no experience and, amazingly, will forge ahead with the job without doing any research or attempting to gain any education whatsoever about slate roof installation. This is despite the fact that a couple of hours on the Internet will yield a wealth of information about slate roofs. Much of that information is available at TraditionalRoofing.com, but there is also much more at SlateRoofCentral.com. If a contractor doesn't understand the principle of headlap, he or she should not be installing a slate roof until that simple principle is completely digested and absorbed.

Let's start with the basics. Every piece of roofing slate has a front and back, top, and bottom, when being used on a standard American-style slate roof. The bottom front (often called the "face") is what you see after the slate has been installed on the roof. The back of every slate is positioned toward the roof (except the starter course, which is installed back side up). The back does not show the beveled edges of the slate nor does it have indented (countersunk) nail holes. The front of every slate faces the sky and shows the beveled edges and indented nail holes.

The top of the slate is covered by the overlying slates. The very top of the slate is called the "head." When one slate overlaps another, that's called overlap. When the head of a slate is overlapped by the slate two courses above it, that's called the headlap — that's the critical overlap on a slate roof. Headlap is what keeps the water from penetrating through the roof. Headlap is essentially what renders the roof watertight. If there is no headlap, the roof will leak.

Standard headlap these days is 3"; however, the headlap varies according to slope. When the slope drops, water drains more slowly, so more watertightness is required. This is achieved by simply increasing the headlap to 4" or even more. Slate roofs should not be installed below a slope of 4:12 (4 feet of rise on 12 feet of run). In fact, the best slate roofs are installed on slopes too steep to walk upon (8:12 or above). However, if you must install slate on a lower slope, then from 4:12 up to 8:12, a 4" headlap is recommended. 8:12 and above can use a 3" headlap, which can even be dropped to 2" when the roof is really steep. These are standard industry recommendations.

Some contractors think 2" headlap is enough. The problem with installing slate with 2" headlap is that the layout of the slate courses requires precision in order to maintain a consistent 2" headlap throughout the field of the roof. If the slates are not installed along a chalk line, the courses can go wavy and you can lose your headlap altogether. Better to install the roof with a 3" or 4" headlap and not worry about cutting corners.

One reason contractors do cut corners with headlap is because less headlap requires less slate, which also reduces the weight of the roof. Neither of these two factors is justification for installing a roof with inadequate headlap. Better to err on the side of caution and install a roof with extra headlap than skimp and risk having to remove the entire roof and start over.

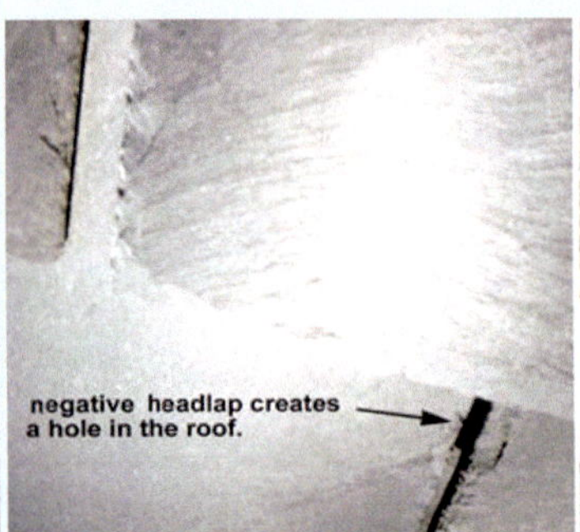

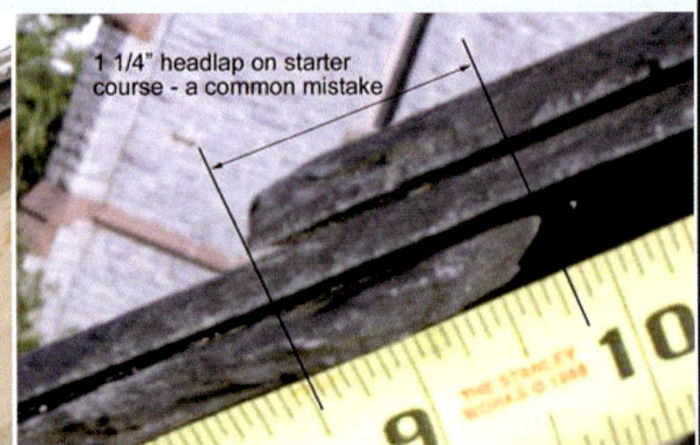

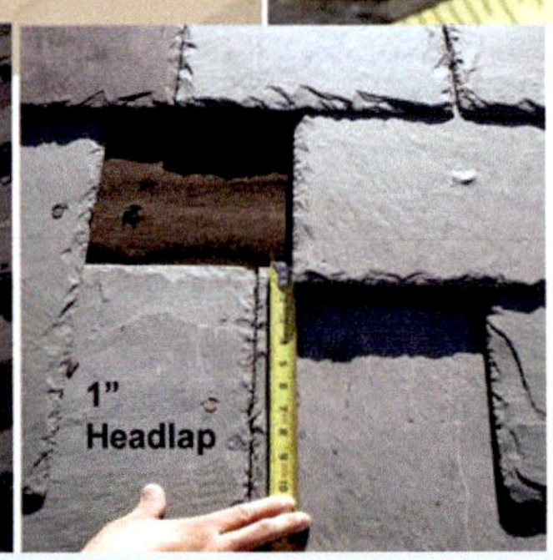

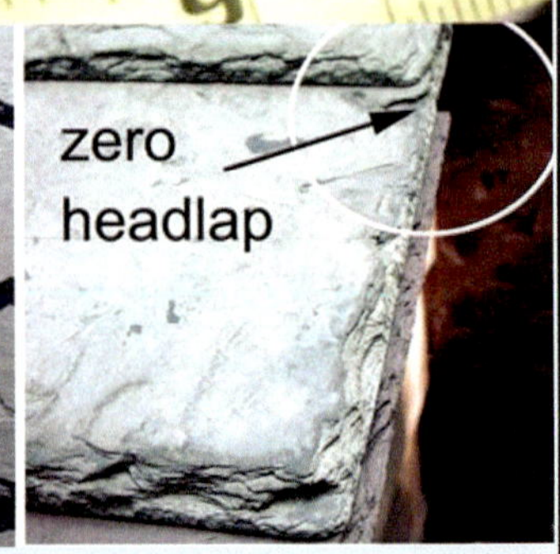

[Headlap, continued]

A roof installed with negative headlap is like a sieve, inadvertently designed to allow water to pass through it. This happens on new slate roofs as big as college dormitories. Headlap is hard, if not impossible, to spot on a roof when looking straight at the field of the roof, especially from the ground. However, if you can look at the gable end, the headlap, or lack thereof, is as clear as day.

There are times when extra headlap is valuable. These include when slate is installed in ice-dam-prone areas of the roof, such as along eaves. If ice damming is a serious consideration on your project, increase the headlap along the eaves by an inch or two when installing the slate. Extra headlap can also help with wind resistance. The extra overlapping makes for a tighter roof, one less likely for wind to get underneath the slates.

The roof calculations are simple enough. If you subtract the headlap measurement from the length of your slate, then divide the remainder in two, you have your slate exposure or course spacing. For example, a 20" long slate with a 4" headlap would have an 8" course spacing (20" minus 4"=16"/2=8"). Course spacing, or *exposure*, is what you see of each course on the roof. Typically, a slate roof is installed by chalking the top of each and every slate course, then nailing the slate to the roof along the chalk lines. In the example given above, there would be a chalk line every 8" up the field of the roof. The top, or head, of the slate would align with the chalk line. The second course of slates above would then overlap that head by 4". Before you run out and start chalking 8" lines up a roof for 20" slates, let me remind you that the starter course, first course, and often second course will be chalked on a different measurement (see page 208).

When ordering slates for a project, you should know your headlap ahead of time and order the slates accordingly. For example, a 10"x20" slate requires 170 slates per square (a "square" is 100 square feet of roof coverage) when installed with a 3" headlap, but requires 180 slates per square when installed with a 4" headlap. Furthermore, you want to make sure the slates are manufactured with the nail holes in the right place. The nails should be installed just above the head of the underlying slate. In the example given above, the nail holes must be a minimum of 12" from the bottom of the slate (8" exposure plus 4" headlap), but 13" would be better because it would give you some extra clearance. If the nail holes are too low, you'll be nailing through the head of the underlying slates, and that's a mistake. You may want to install a slate roof with even more than 4" of headlap on a lower slope situation, so you'll have to watch your nail placement; most slates are manufactured to be installed with up to a 4" headlap.

It's not as confusing as it sounds. Study the illustrations here and you will see that headlap is really a simple concept as well as a critical element of a slate roof. You will also find that, once you understand headlap, you can make a roof out of anything that can be made into a flat shingle, not just a piece of stone.

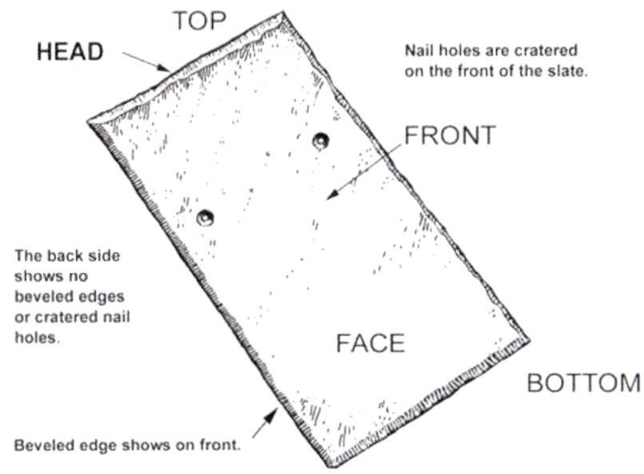

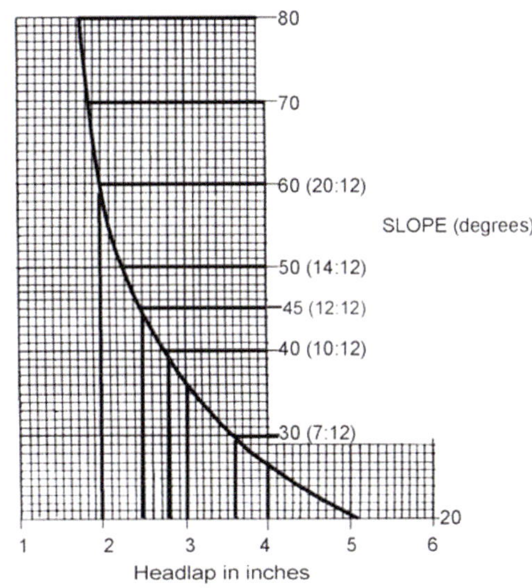

Headlap Variations with Change of Slope

Source: *Slating and Tiling* by J. Millar. English Universities Press Ltd. 2nd ed. 1947. p.34

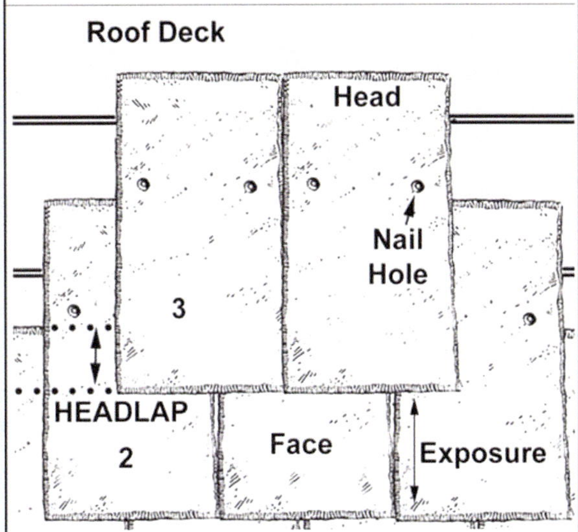

The head of the slate in course 1 is overlapped by the face of the slate in course 3. This is called headlap.

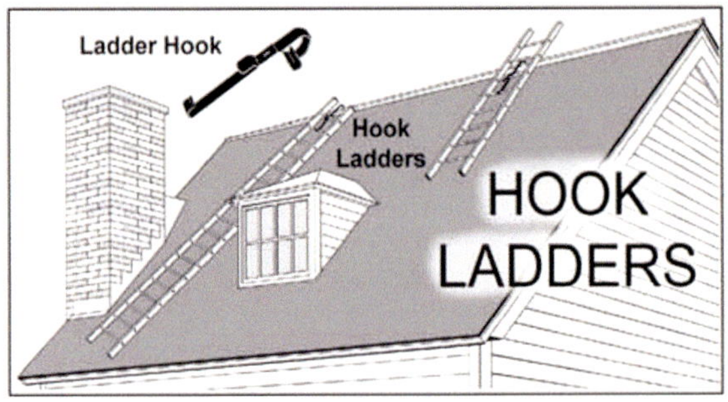

Hook ladders are angled on the roof when needed (above). The hooks can be installed on any rungs.

Ladder hooks are available with and without wheels and can be used single or double (above).

Roof jacks can be nailed on top of the slates, or slid underneath them, or a slate can be taken out and the roof jack nailed between the underlying slates.

How to Replace a Broken Slate

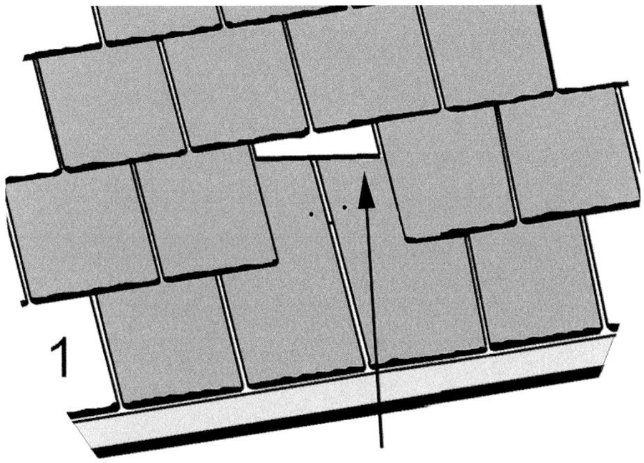

Broken slate needs to be replaced.

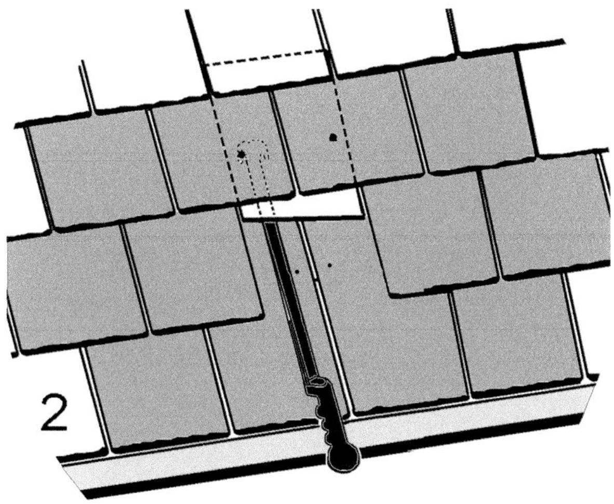

Slate ripper hooks slate nails and pulls them out, removing slate.

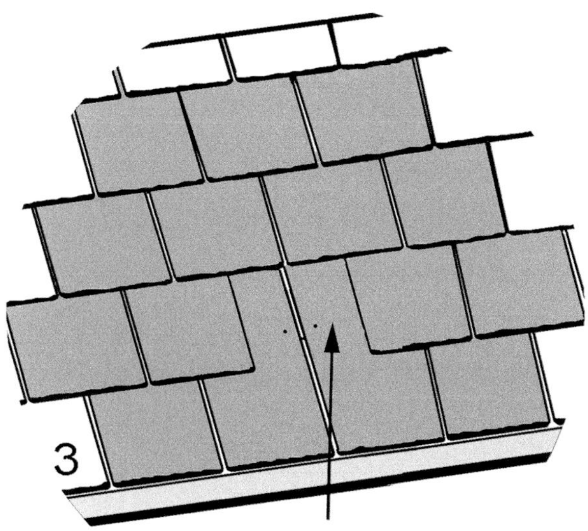

The slate has been removed and the roof is now ready for a replacement.

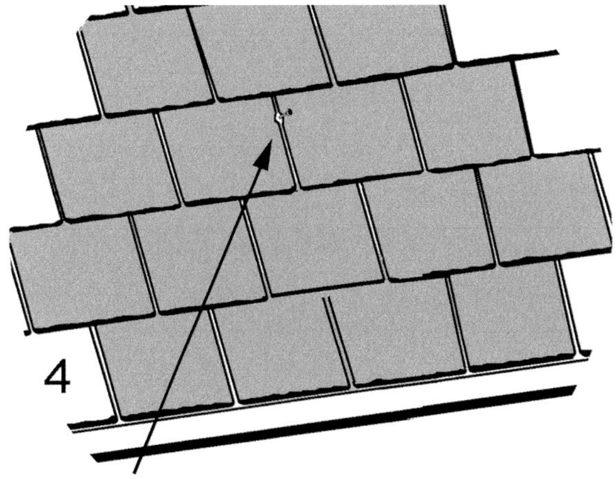

Replacement slate is slid into place and fastened with a nail in the slot.

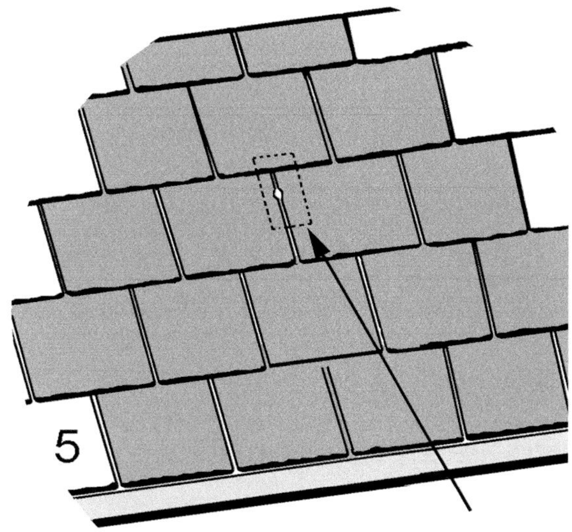

The roofing nail is tapped down and a bib flashing is slid under the slate but over the nail head.

SLATE HOOK REPAIR

Use a stainless steel or copper slate hook instead of a nail and bib. Pound it in, slide the replacement slate over it and hook it into place.

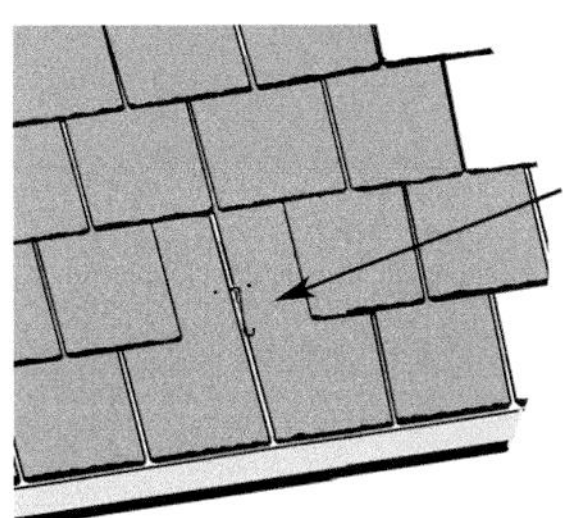

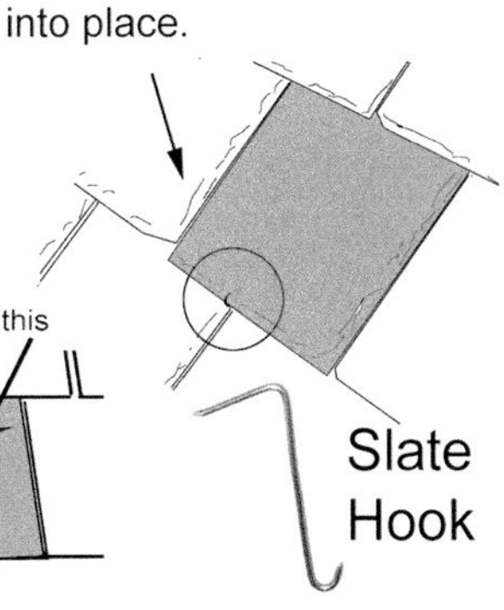

DON'T STRAP REPAIR

NOTES